N. M. Langdon

Nov 1 1877.

Chester, N.J.

A PRACTICAL MANUAL

OF

CHEMICAL ANALYSIS AND ASSAYING,

AS APPLIED TO THE

MANUFACTURE OF IRON FROM ITS ORES, AND TO
CAST IRON, WROUGHT IRON, AND STEEL,
AS FOUND IN COMMERCE.

BY

L. L. DE KONINCK, Dr. Sc., and E. DIETZ,

ENGINEERS.

EDITED, WITH NOTES,

BY

ROBERT MALLET, F.R.S., F.G.S., M.I.C.E., Etc.

FIRST AMERICAN EDITION,

EDITED, WITH NOTES AND AN APPENDIX ON IRON ORES,

By A. A. FESQUET,

CHEMIST AND ENGINEER.

PHILADELPHIA:

HENRY CAREY BAIRD,
INDUSTRIAL PUBLISHER,
406 Walnut Street.
1873.

PHILADELPHIA :
COLLINS, PRINTER, 705 JAYNE STREET.

PREFACE OF THE AMERICAN EDITOR.

This Manual is designed for practical use in the iron master's laboratory. Its intent and scope have been fully set forth in the Preface of the Authors and that of the English Editor, so that we need not here enlarge upon the subject.

The American editor has preserved intact the text of the authors and the notes of the English editor, and has added besides various additional notes and an Appendix, in which latter the principal Iron Ores are examined in a general manner, in regard to their composition, quality, and geological disposition. Instead of placing the notes at the end of the book, and referring to them by means of numbers, they have been intercalated in the text, at the places to which they belong. The American editor's notes are placed between brackets.

The vast and steadily growing importance of the iron manufacture in the United States, and the increasing interest in the chemical researches relating to it, induce us to hope that this Manual **may** help to popularize the study of chemistry, to which of late years the metallurgy of iron owes its greatest progress.

A. A. F.

PHILADELPHIA, September 10, 1873.

ENGLISH EDITOR'S PREFACE.

This small volume deals simply with the practical. It is the work of authors skilful and well acquainted with the analytical methods generally adopted in the great ironworks and factories of Belgium, France, and Germany, which, together with the apparatus and reagents there usually employed, they succinctly describe.

The work appeared to me a useful table manual, even to the accomplished assayer and analyst. It is also one, from the careful study of which, accompanied by the self-instruction derivable from a repetitive course of the operations described, any tolerably intelligent man, with some preliminary knowledge of inorganic chemistry and of manipulation, might become a practical iron-assayer. As much chemistry as that may now be acquired at many of our educational institutions, colleges, etc. Although our national notions and standard of

1*

general education remain so defective, that engineers, mechanicians, founders, manufacturers, and traders remain generally ignorant of chemistry and chemical analysis, the knowledge of which is yet so important to all, a clearly written handbook such as this (which deals with the vastest of our metallic industries— that of iron) cannot but prove serviceable now, and must become increasingly so with the progress of education.

I have, therefore, deemed the work of Drs. De Koninck and Dietz worthy of translation, and have added some notes which, it may be hoped, do not detract from its value.

Amongst these is one which may offer several useful hints (derived from personal experience) for the construction and arrangement of industrial and analytical laboratories in ironworks and like establishments. The formulæ, as well as the atomic and molecular numbers, I have left as I found them in the original work.

ROBERT MALLET.

LONDON, 3d June, 1872.

PREFACE OF THE AUTHORS.

AMONG the numerous works of analytical chemistry which we have consulted, we have not met with a single treatise on the docimacy of iron in an exclusively industrial point of view. We have tried to supply this want by the publication of the present Manual, in which our aim has been to present in a concise form every item of information which we considered of use to chemists in ironworks, for whom our work is specially intended.

We are convinced that our labor will spare them long and wearisome researches, which they would often be obliged to make, through more general treatises often of little practical utility. We hope we shall assist in generalizing the processes most in use, or at least in causing the adoption of uniform methods of operation, the introduction of which would be especially valuable in cases of conflicting analyses.

If we succeed in our endeavor, we shall be sufficiently recompensed for the labor of the investigations we have undertaken.

LIEGE, 1871.

CONTENTS.

PART I.

THE REAGENTS.

PART II.

APPARATUS AND OPERATIONS.

PART III.

VOLUMETRIC ANALYSIS OF IRON.

PART IV.

ANALYSIS OF IRON ORES.

  2

PART V.

ASSAY OF IRON ORES BY THE DRY METHOD.

PART VI.

ANALYSIS OF CAST-IRON, MALLEABLE IRON, AND STEEL.

PART VII.

ASSAY OF FUELS.

SUPPLEMENTARY NOTES BY THE AUTHORS.

APPENDIX.

BY THE AMERICAN EDITOR.

A PRACTICAL MANUAL

OF

CHEMICAL ANALYSIS AND ASSAYING.

PART I.

THE REAGENTS.

I.—SIMPLE SOLVENTS.

Water.

Formula, H^2O.   Atomic weight, 18.

IF water containing even a slight trace of impurity were employed for analytical purposes, we should be liable to fall into serious error, taking into consideration the large quantity of water required.   Only distilled water is employed.   For the necessary degree of purity we should ascertain that there be no residuum after evaporation, and that its clearness be preserved after the addition of a barytic solution or the oxalate of ammonium; any disturbance in the water occasioned by the first of these reagents would indicate the presence of sulphuric acid or some sulphate; the second would indicate, in like manner, the presence of salts of calcium.   A small quantity

2

of the sulphhydrate of ammonium should only give water a clear yellow tint; a greenish tint or a precipitate would be produced by the metallic salts. Lastly, we should ascertain that water acidulated with pure sulphuric acid does not discolor permanganate of potassium, even at a temperature of 60° to 70° Centigrade, which would indicate the presence of organic matters, and would injure the analysis of iron by the volumetric process of Marguerite.

The presence of small quantities of hydrochloric acid, or of any alkaline chloride, will not falsify the results of such analyses as we are here engaged with; a solution of nitrate of silver will give a very simple and sensitive test for their presence. Water should be distilled in a copper still, coated internally with tin, and furnished with a tin worm or condensing tube. The first water distilled over is always rejected, as it generally contains ammonia, and we should not distil off more than two-thirds of the water employed in order to avoid the carrying over of solid matters. In large manufactories distilled water is easily obtained by condensing the vapor issuing from a steam boiler.

The editor writes from his experience of some years in his own laboratory, which for some years was in the midst of large engineering works, when he states that, whilst it is quite possible to procure an ample and constant supply of quite pure distilled water from steam taken off from the boilers of factory engines, *by the employment of proper precautions*, it is quite impossible to obtain it except in a more or less impure state by the methods usually employed for taking the steam off and for condensing it. As steam comes rushing through the large steam pipes from the boilers to the engines, it brings along with it sometimes mechanically-carried earthy matters, or even common salt carried up with the steam, but always organic matters from the oil, tallow, packings, etc., and occasionally sensible traces of copper and iron, etc. The impurity always to be guarded against, however, is greasy organic matter, which, small as it is in quantity, may cause chemical obscurities, and produce a film of grease upon the glass apparatus disagreeable and hard to clear away.

The editor found that the *best* method to obtain absolutely pure water, was to employ the heat of the engine steam to redistil water previously condensed from the steam boiler, from a thin silver alembic, the water being maintained therein at a nearly constant level by a simple float arrangement (the converse of that commonly employed to discharge condensed water from steam pipes), and to condense the vapor from the alembic in a long silver tube enveloped in a current of cold water. But water sufficiently pure for almost any laboratory purpose can be obtained directly from the usual sorts of steam engine boilers by the following precautions : The steam should be taken off as far as possible from the boiler itself, and by a rather small tube from the steam pipe, above which it should rise vertically for some feet, and there pass through a large, well-burnt stoneware pipe filled full of quartz pebbles, carefully washed in dilute hydrochloric acid, and afterwards exposed to boiling water for some time.

If the *surface* be sufficient in such a tube of pebbles (not bigger than large peas) the whole of the organic and mechanically-carried-up matters (not soluble) are separated by contact action ; and if the steam boiler be fed with reasonably pure natural water, and there be no copper or brass in stop valves, etc., between the boiler and the place of taking off the steam, scarcely anything can be detected in the distilled water after the apparatus has been some days at work. The condensation was effected by the editor simply by passing the steam from the stoneware tube vertically upwards into a large German glass tube immersed externally in a constant run of cold water. The store of distilled water was kept in stoneware jars of the largest size, with ground-glass stoppers. Thence for immediate use it was transferred to a globular bottle of (bright metal) crown glass, with a ground stopper, the bottle, which held about a gallon, resting upon a small balanced stand, on the laboratory table, so that it could be tilted easily, to draw off water.

Cocks, however made, are objectionable ; the moist uncovered surfaces collect dust and other impurities. Flint-glass bottles are inadmissible for holding pure water.

[Worms made of tin for condensing distilled water will be found to answer the purpose well, provided the metal employed be pure block tin, and not, as is often the case, an alloy of tin and lead.]

To keep water pure, we should preserve it in glass bottles from the vapors of the laboratory, and never pour it directly from these into vessels filled with vapors or gases. A very suitable apparatus for the preservation of water consists of a large bottle (Fig. 1),

Fig. 1.

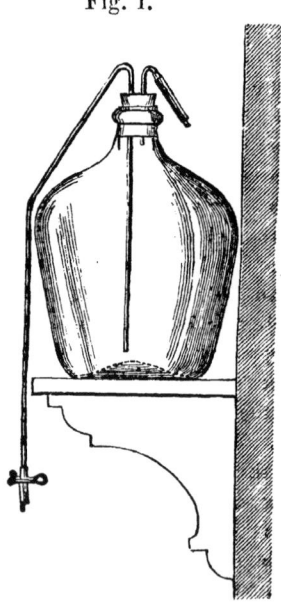

furnished with a doubly perforated cork, having on one side a siphon, and on the other a tube for the access of air; the siphon terminates in a gutta-percha tube closed with a stopcock; the air tube is bent down, and contains a bung of wadding to retain dust. It would be still better to cause it to communicate with two small vessels, one containing sulphuric acid and the other a solution of potassa or caustic soda, in order to prevent the gases or injurious vapors from the laboratory penetrating the apparatus. This can be placed on a stand or bench of suitable height. For ordinary use, water can be preserved in phials with a spout or tube.

Alcohol.

Formula, C^2H^6O.   Atomic weight, 46.

The alcohol employed is not that represented by the above formula; ordinary spirits of wine at 85 per cent. is quite suitable.

Ether.

Formula, $C^4H^{10}O$.   Atomic weight, 74.

The ether of commerce is sufficient for general purposes. Even at an ordinary temperature, ether emits vapors which form an explosive mixture with air. Care should be taken not to make use of this reagent in proximity to any body in combustion.

Serious accidents have originated in the blowing out of the stoppers, or by the bursting of large bottles of ether, exposed to much radiant or other heat in laboratories.

The store bottle of ether is best kept within a cylindrical tin canister, and in a part of the laboratory as remote as possible from furnaces, gas flames, or direct sunshine, etc.

II.—ELEMENTARY BODIES.

Hydrogen.

Formula, H.   Atomic weight, 1.

Hydrogen is prepared by the action of metallic zinc, and dilute sulphuric acid. The following equation shows the reaction : $Zn + H^2SO^4 = ZnSO^4 + H^2$.

To obtain pure hydrogen the zinc and the acid used should both be pure, and the gas should not be collected until the air is completely expelled from the apparatus. The latter is composed of a bottle with a wide neck, closed with a doubly perforated cork, supporting a funnel tube extending to the bottom of the bottle, and a bent tube for the liberation of the gas. The bottle contains water and granulated zinc, and upon this we pour through the funnel some sulphuric acid, adding more from time to time to secure the regular development of the gas. We may also use the apparatus of Doebereiner, described on page 32, in reference to the preparation of carbonic anhydride. To obtain dry hydrogen we must pass it through a tube of the U form containing chloride of calcium.

Belgian zinc is much purer than British or German, and "punchings" out of Belgian sheet zinc, in a perfectly clean state can be procured in London from the zinc workers, which are very convenient for the evolution of hydrogen.

Perfectly pure hydrogen, however, cannot be obtained at once from any zinc of commerce, and if so evolved, it should be passed through a strong solution of caustic potassa, and then through one of bichloride of mercury, if needed pure.

Pure hydrogen may be evolved from sodium, or by the galvanic decomposition of water between platinum electrodes, where the volume required is not large.

[American zinc is purer than the European article, especially in its freedom from arsenic and lead. It often contains, however, a certain amount of iron, when remelted in cast-iron kettles preparatory to pouring it into moulds. We have used for analytical purpose a sample of zinc made at Newark, N. J., which was entirely free from arsenic and iron.]

Chlorine.

Formula, Cl. Atomic weight, 35.5.

Chlorine gas is obtained by taking away the hydrogen from hydrochloric acid, by means of an oxidizing body. The peroxide of manganese is the most suitable of all proposed oxidizers.

The apparatus required is represented in Fig. 2 ;

Fig. 2.

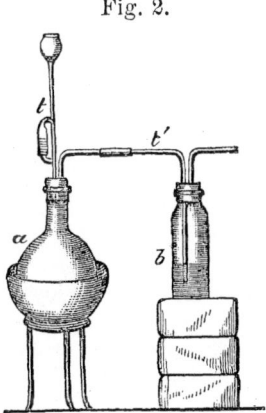

a is a flask containing a mixture of the peroxide and ordinary hydrochloric acid ; it has a safety-tube t, and a discharge tube t'. The latter conveys the gas into the flask b, containing a small quantity of water. If it be desirable to obtain dry gas, a second flask containing concentrated sulphuric acid is added to the apparatus. The flask a is half dipped in a water-bath. The reaction commenced then goes on in the cold, but is accelerated by the application of gentle heat. The peroxide of manganese, in pieces of the diameter of one to two cubic centimetres (size of large beans), is preferable to the powdered oxide, not only from its greater purity, but from the more regular liberation of the chlorine. The following equation represents the reaction:—

$$8HCl + 2MnO^2 = Mn^2Cl^4 + 4H^2O + Cl^4.$$

Chlorine water is obtained by passing the gas through distilled water up to saturation; this solution, which contains about five volumes of chlorine to one of water, must be kept in a cool place, excluded from light, in a glass stoppered bottle, rendered air-tight by keeping it turned upside down, to make by the fluid itself the closure more hermetical. Notwithstanding these precautions it changes before long; the chlorine combines with the hydrogen of the water (forming hydrochloric acid), and oxygen is liberated.

Tin cylindrical canisters with covers, or those of hard wood turned hollow, with caps to slip on, are used in all the German laboratories to preserve preparations acted on by light, and are much more convenient than shutting the bottles up in cupboards and drawers. Bottles of the opaque black glass, furnished by some of the chemical apparatus dealers, are objectionable, as it is impossible to judge of the state of their interior.

Bromine.

Formula, Br. Atomic weight, 80.

Iodine.

Formula, I. Atomic weight, 127.

These two reagents are employed in the state furnished by commerce. Bromine is preserved in well-stoppered and capped bottles, as a security against escape of its vapor, which is excessively corrosive. In using this reagent we should not forget that it has a most injurious effect upon the skin, and that its fumes attack the eyes and lungs. Bromine water is prepared by shaking together briskly in a closed vessel bromine and distilled water ; to insure saturation, the bromine should be added until some remains undissolved. We can ascertain whether the solution be free from sulphuric acid by means of chloride of barium.

Bromine is employed in large quantities for the determination of carbon in cast iron ; and as it is expensive, it is advantageous to preserve the solution obtained in order to extract the bromine again. In the solution it is found in the state of bromide of iron ; with the aid of a solution of carbonate of potassium, we effect a double decomposition, and obtain an insoluble compound of iron, and a soluble one of bromide of potassium ; by filtration and evaporation we obtain crystals of the latter. To abstract the bromine, we mix these crystals with pulverized peroxide of manganese and sulphuric acid, diluted with three-fourths of its volume of water. This mixture is then distilled, and the bromine vapors condensed in a small glass refrigerator tube.

Iodine is recovered in the same way, taking care to substitute concentrated sulphuric acid for the dilute.

At existing prices in commerce it is scarcely worth while, in any British laboratory, to attempt the disagreeable process of recovering either bromine or iodine residues. They are best bargained for at their value with the chemical dealer who supplies the laboratory.

Bromine, as has been shown by Dr. Waage, may with advantage be employed as a general oxidizing agent in the metallurgical laboratory. It may be employed in three forms—as free bromine ; as a solution in water, which, at ordinary temperatures, does not contain more than 3 per cent. of bromine ; and in solution in strong hydrochloric acid, which dissolves about 12 per cent. Sulphides, even iron pyrites in crystals, are readily decomposed by bromine. Sulphur is by it more rapidly oxidized than by strong nitric acid ; and precipitated sulphides are thus easily broken up, and brought to a state fit for weighing, without the necessity of burning the filter. The presence of ammoniacal salts (with which bromine liberates nitrogen) hinders the formation of peroxides in acid solutions of cobalt, nickel, and manganese, but does not interfere with that in the like solutions of iron, tin, and mercury. (Fresenius, Zeitschrift, 1872.) The advantages of bromine water over chlorine water as an oxydant are considerable. According to the writer's observation, bromine and iodine together acts more energetically in breaking up cast iron, for liberation of its carbon, than either alone ; and bromine, through which chlorine has been passed, acts more rapidly than that element alone.

Oxygen.

Formula, O. Atomic weight, 16.

Oxygen is obtained pure by the decomposition of chlorate of potassium by means of heat :—

$$KClO^3 = KCl + O^3.$$

The operation is performed in the following manner: Pour some pulverized chlorate of potassium, mixed with about a fifth of its weight of peroxide of copper or pulverized peroxide of manganese, into a small glass retort. The oxides of copper or manganese greatly assist the reaction, although they are not themselves decomposed. Care should be taken that the mixture contains no organic matter, which might cause explosion, or at least destroy the purity of the gas. An escape-tube is fitted to the retort by means of a cork or caoutchouc tube, and heat is applied by a spirit or gas lamp or a little charcoal furnace. When the liberation of the oxygen has commenced we collect the gas in a gas vessel, by inserting the escape-tube in the tubulure of that vessel. The temperature should be raised towards the end of the operation, but carefully, to avoid bumping at the bottom of the retort. The proportion of chlorate to be employed is four grammes for each litre of oxygen required. The peroxide of copper made use of is not lost; to regain it we have only to add water to the residuum of the operation, and by filtration separate the insoluble oxide, and then calcine it afresh.

[The safest way to prevent a too rapid disengagement of gas, is to considerably dilute the chlorate of potassium, as indicated in the text, by the peroxides of copper or of manganese, or to add only a very minute proportion of peroxide of manganese, one-thousandth part for instance.

Care should be taken to ascertain that the peroxide of manganese contains no sulphuret, otherwise violent explosions may occur.

If a caoutchouc (India-rubber) tube be employed for the apparatus, it should be put where the heat is not strong enough to affect it. By the decomposition of India-rubber by heat, hydrocarbons are formed, which produce explosive mixtures with oxygen, or contaminate the purity of this gas for analytical researches.]

Iron.

Formula, Fe. Atomic weight, 56.

Iron is made use of to fix the standard strength of the solutions intended for volumetric analysis. For this purpose we use piano-forte *wire*, which is the purest iron furnished by commerce ; it contains on the average $\frac{4}{1000}$ impurities, and consequently $\frac{996}{1000}$ pure iron.

It should be preserved free from damp, and from the fumes of the laboratory, which would cause oxidation. It may be kept in the balance room and best in a wide stoppered bottle.

Zinc.

Formula, Zn. Atomic weight, 62.2.

Thin plates of rolled zinc are most generally employed : it is cut into small pieces of equal size, which facilitates the operation, as we shall see when treating of the volumetric determination of iron. The rolled metal of commerce is not of sufficient purity to standardize solutions of sulphide of sodium intended for the volumetric analysis of zinc; it is necessary to employ for this redistilled zinc.

Tin.

Formula, Sn. Atomic weight, 118.

This metal is employed in thin slips for the determination of phosphoric acid. For the preparation of chloride of tin, grain tin of commerce may be employed.

III.—ACIDS.

Hydrochloric Acid.

Formula, HCl.   Atomic weight, 36 5.

The pure acid is colorless, and leaves no precipitate when evaporated in a capsule of platinum; it ought not to turn blue immediately on the addition of a little starch paste and iodide of potassium, which would indicate the presence of free chlorine or ferric · chloride.   Diluted with twice or thrice its volume of water, it should not be affected by a drop of a solution of chloride of barium, as indicating the existence of sulphuric acid.   Besides pure hydrochloric acid, we use for the preparation of chlorine, etc., the impure acid sold under the name of ordinary hydrochloric acid, which is strongly colored yellow by ferric chloride.

Nitric Acid.

Formula, HNO^3.   Atomic weight, 63.

This reagent is employed in two conditions; that of concentrated nitric acid, usually colored yellow or brown by less oxidized combinations of nitrogen, or ordinary nitric acid less concentrated than the first.

This acid ought not to leave any residue on evaporation; it ought to be quite free from sulphuric acid, which can be ascertained as indicated in reference to hydrochloric acid; with a solution of nitrate of silver we can ascertain the presence of chlorine.   A small quantity of the latter is not generally injurious in iron analysis.

3

Aqua Regia.

This is obtained by the mixture of one volume of nitric acid to three or four volumes of hydrochloric acid ; it is not usual to prepare it beforehand, for according to circumstances there is an advantage in making use of aqua regia containing an excess of one or other of these acids.

Sulphuric Acid.

Formula, H^2SO^4. Atomic weight, 98.

When pure this acid is colorless, and when diluted with twice or thrice its volume of water ought not to decolorate a drop of a solution of permanganate of potassium, either immediately, which would be due to sulphurous acid, or after contact for some time with a slip of pure zinc, which would indicate the presence of nitric acid.

A current of hydrosulphuric acid, even when prolonged for some time, should produce no precipitate in the pure dilute acid.

Dupasquier some years ago (Comptes Rendus, t. xx. 1845) proposed a very simple and elegant method of depriving the sulphuric acid of commerce of its most mischievous impurity—namely, arsenic, which exists in it as arsenic acid to the amount of .001 to .0015, according to Dupasquier. His method consists in adding to the impure acid a sufficient amount of sulphuret of barium to precipitate the arsenic ; any excess of the barium is precipitated as insoluble sulphate. This method is useful as enabling the acid of commerce to be used for many laboratory purposes, owing to the slight solubility of sulphate of barium. However, it is at least doubtful if this method leaves an acid absolutely pure.

Hydrosulphuric Acid (Sulphuretted Hydrogen).

Formula, H²S.   Atomic weight, 34.

Hydrosulphuric acid gas is obtained by the action of sulphuric acid or dilute chlorhydric acid on sulphide of iron :—

$$Fe^2S^2 + 2H^2SO^4 = 2H^2S + Fe^2\bar{S}^2O^8 ;$$

or .

$$Fe^2S^2 + 4HCl = 2H^2S + Fe^2Cl^4.$$

Hydrochloric acid is preferable to sulphuric acid, because the latter, by producing a salt easy of crystallization involves a more frequent cleansing of the apparatus.   To prepare this acid introduce into a flask *a* (Fig. 3) some fragments of sulphide of iron and

Fig. 3.

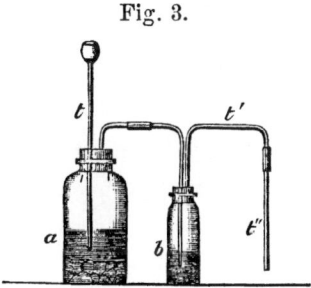

some water : by the funnel *t* pour in the acid as required, and according to the desired rapidity of the current of gas.   The gas first passes into the flask *b*, which contains a little water, thence by the bent tube *t'*, and the continuing tube *t''*, into the liquid upon which it is to act.   This continuing tube, supported by a caoutchouc tube, is easily removed and cleaned.   To prepare a solution of hydrosulphuric acid, we must pass the gas to saturation into distilled water; this solution must be preserved from air and light.

We can very easily prepare sulphide of iron for this use by mixing three parts of iron filings with two parts of flowers of sulphur, projected gradually into an earthern crucible heated red hot. It is covered and heated until the mass is well melted, and then poured out on a metal plate, and broken up.

Sulphuret, or sulphide of iron, is so readily procured in commerce now that it is never worth while to prepare it in the laboratory.

For those who may be so circumstanced, however, as to be obliged to prepare it, the best and easiest method is to heat the extremity of a flat bar of iron (of about $\frac{1}{2}$ in. by 2 in.) *white* hot, and bring it into contact with a piece of roll sulphur ; the iron combines readily, and the fused sulphuret which drops off may be received in a vessel of water.

Acetic Acid.

Formula, $C^2H^4O^2$.　Molecular weight, 60.

Ordinary acetic acid, containing about thirty per cent. of the normal acid is used; it is sufficiently pure when it leaves no residuum upon evaporation.

Oxalic Acid.

Formula, $C^2H^2O^4 + 2H^2O$.　Molecular weight, 126.

This acid is employed to standardize solutions of permanganate of potassium. The commercial acid is never sufficiently pure for the purpose; it must be purified by recrystallization. To do this we must pour such a quantity of hot distilled water on the commercial acid as will not completely dissolve it, then filter hot and leave the liquid at rest. In cooling, crystals are deposited which are separable by decanting the mother liquor. These are left to dry between sheets of filtering paper. The drying must be done without heat, in order to avoid driving off some of the water of crystallization.

This acid is pure when no residuum is left after combustion on platinum foil.

[Certain kinds of commercial oxalic acid are wet from the presence of nitric acid. They should be washed first upon a filter, with cold water, in order to remove the foreign acid, and then made to recrystallize.

Dr. Mohr recommends the use of tepid, instead of boiling, water, added in quantity insufficient to dissolve the whole of the oxalic acid under treatment. The impurities, composed mostly of the acid oxalate of potassium, are but slightly soluble in a saturated solution of oxalic acid.]

Succinic Acid.

Formula, $C^4H^6O^4$.   Molecular weight, 118.

This is only required for the preparation of the succinate of ammonium (see this reagent). To ascertain its purity, observe whether there be any residuum after combustion on platinum foil. It is preserved in the state of crystals which should be colorless.

[### Citric Acid.

Formula, $C^6H^8O^7$.   Molecular weight, 192.

This acid is generally prepared from the juice of lemons. Like tartaric acid, its principal use in the analytical laboratory is to prevent the precipitation of certain substances in solution, iron and alumina, for instance, by alkalies. Citric acid is recommended by certain persons, as superior to tartaric acid for such purposes. Its solution in water decomposes and becomes mouldy

3*

after a certain length of time. It is therefore better kept in the pulverized form, buying it, however, in the shape of crystals, which should be as dry as possible. It is often contaminated with sulphuric acid, which, in many analytical cases, may be passed over, but citric acid should leave no residuum after combustion on platinum foil.]

Tartaric Acid.

Formula, $C^4H^6O^6$. Molecular weight, 150.

In solution in pure water, tartaric acid decomposes rapidly, and becomes mouldy. The solution can be preserved unadulterated by the introduction of a morsel of camphor. The acid may be preserved in a solid state, best in fine powder, so that a solution can be rapidly made when required. The same test as that for the preceding acid decides its purity.

If a solution of tartaric, or even of citric acid, which is still more readily alterable, be made with boiling distilled water, free from organic matter, and the hot solution without filtration decanted into a stoppered bottle, it remains good and free from mould for a very long period.

To acids properly so called we shall add the three following *anhydrides :—*

Sulphurous Anhydride.

Formula, SO^2. Molecular weight, 64.

Sulphurous anhydride is produced by deoxidizing sulphuric acid by means of metallic copper: the following equation indicates the reaction :—

$$2H^2SO^4 + Cu = CuSO^4 + 2H^2O + SO^2.$$

The apparatus required is the same as that used for the preparation of chlorine, with this difference only—it is heated by a naked charcoal fire or by gas-flame ; into a flask a (Fig. 2) pour four parts of ordinary sulphuric acid, then add one part of copper wire or sheet cut into bent shreds ; fill the tube intended to prevent absorption, also with sulphuric acid. The heat required must be cautiously applied to avoid bumping or bubbling over of the mass. From the flask, b, the gas is conveyed either into the liquid upon which it is intended to act, or into a flask containing cold distilled water, and kept cold, if a solution be required. The solution must be preserved in well-closed bottles. Charcoal may be used instead of copper; the operation is performed in the same way, but the following equation, $2H^2SO^4 + C = 2H^2O + CO^2 + 2SO^2$, shows that the sulphurous anhydride thus obtained is mixed with half its volume of carbonic anhydride, which much impairs its solubility, and consequently its action.

Carbonic Anhydride.

Formula, CO^2. Molecular weight, 44.

To prepare this gas, drop into a flask of water some pieces of carbonate of calcium ; the cork of the flask should be perforated by an escape-tube and a funnel-tube, descending into the liquid; through the latter tube, pour in gradually some hydrochloric acid, regulating the quantity by the rate of liberation of the gas produced.

The reaction is as follows :—

$$CaCO^3 + 2HCl = CaCl^2 + H^2O + CO^2.$$

The apparatus of Doebereiner, represented in Fig. 4, is very suitable for the preparation of this gas.

In the middle of a cylindrical vessel, containing HCl, a glass bell with tubes is supported by means

Fig. 4.

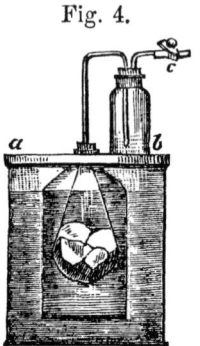

of a disk of wood or metal. Some carbonate of calcium (white marble) in tolerably large pieces is suspended in this bell by means of a wire net, or is placed in a perforated vessel placed within the bell. On letting the gas escape from the bell, the dilute hydrochloric acid ascends in it, and coming in contact with the carbonate of calcium, evolves carbonic anhydride, which escapes by the tube passing through the flask, b, placed on the top of the apparatus. On the bell being again occupied by the gas the hydrochloric acid retreats, and the apparatus is at rest. It is easy to regulate the escape of the gas.

Molybdic Anhydride.

Formula, MoO^3. Molecular weight, 144.

This anhydride is used for the preparation of molybdate of ammonium. The commercial preparation is sufficiently pure.

IV.—BASES.

Potassa.

Formula, KHO.   Molecular weight, 56.

Commerce furnishes two kinds of potassa: lime potassa and alcohol potassa; the first is altogether impure, and should be entirely rejected; even the second often contains impurities. If the potassa be pure, a watery solution, neutralized by hydrochloric acid, should not be affected by hydrosulphide of ammonium; potassa in contact with the air, especially in solution, quickly absorbs CO_2. It should be preserved in the solid state in a well-stoppered bottle.

Soda.

Formula, NaHO.   Molecular weight, 40.

What we have said with regard to potassa, applies equally to soda. The same test for purity as for potassa is required.

Notwithstanding what is here said in the text, it is indispensable to keep caustic potassa and soda, always in solution. The only real difficulty in doing so consists in this, that finely-ground glass-stoppered bottles must be used to contain them, and that strong caustic alkaline solutions act sufficiently rapidly upon *ground* surfaces of glass to cement the stoppers into the bottles, even in a few days' time, so that they cannot be gotten out.

A simple and complete remedy for this is to *thinly* coat the ground neck of the bottle and the stopper, both being dry and warm, with paraffine and hard tallow melted together in equal volumes.

This is not sensibly acted upon : keeps the stopper perfectly air-tight, and prevents its sticking fast. The bottles used should be hard German glass or British (bright metal) crown glass ; and after long standing it is prudent to see if the solutions have taken up any silex from the glass, as well as any carbonic acid from the air.

Ammonia.

Formula, NH^3. Molecular weight, 17.

Ammonia (a watery solution of the gas) should be colorless, and ought to evaporate without leaving any residuum ; it is often important to obtain it free from carbonate ; the test is lime-water, this reagent should not occasion any cloudiness.

Lime.

Formula, CaO. Molecular weight, 56.

Lime-water (solution of the hydrate of calcium) is obtained by digesting slaked lime in excess, with cold, distilled water, filtering and decanting the clear liquid rising to the surface.

The lime employed should be obtained by calcining in a porcelain tube, with free access of air, and a little water-vapor passed through with it, white Carara marble. Few limestones are free from magnesia and silica in a soluble state.

Alumina.

Formula, Al^2O^3. Molecular weight, 105.

This combination is employed as a flux in assays by the dry method. The alumina (pipe-clay) of commerce is generally sufficiently pure for this purpose.

Oxide of Lead.

Formula, PbO. Molecular weight, 223.

Pure litharge free from minium, the presence of which may be known by its reddish color, is employed. It is finely powdered and passed through a sieve; by this means it is freed from metallic lead if it should contain any. It is only employed for dry assaying.

[Dioxide of Lead.

Formula, PbO^2. Molecular weight, 239.

This substance is useful for ascertaining qualitatively the presence of small quantities of manganese in iron ores. The dioxide of lead is obtained by digesting red lead in dilute nitric acid, which dissolves the protoxide of lead, and leaves the dioxide in the shape of a deep brown powder. For analytical purposes this substance should be free from chloride of lead.]

Oxide of Copper.

Formula, CuO. Molecular weight, 79.

The black or peroxide of copper is employed either in a finely granular state, or in powder. It is used for the quantitative determination of carbon by combustion, and consequently should be free from organic matters, from which it is purified by calcination.

V.—ALKALINE SALTS.

A.—Salts of Potassium.

Nitrate of Potassium.

Formula, KNO^3.   Molecular weight, 101.

For analytical use, nitrate of potassium is sufficiently pure when its solution in distilled water is not affected by the addition of a few drops of a barytic solution.   Nitre is used in the dry method as an oxydant.

The nitre of commerce, though well crystallized, frequently contains sodium, as sulphates and chlorides of that, and of potassium.   By two crystallizations, however, it may be rendered almost quite pure.

Sulphate of Potassium.

Formula, $KHSO^4$.   Molecular weight, 136.

The hydrosulphide of ammonium should not occasion any precipitate when poured into a neutralized solution of this reagent.   The formation of a precipitate would indicate the presence of alumina or of metallic salts.

[The formula given above indicates that this substance is what is generally known under the name of *acid sulphate of potassium*, and which is obtained by evaporating in a platinum vessel, nearly to dryness, a mixture of 1 part in weight of the powdered neutral sulphate with $\frac{1}{2}$ part of oil of vitriol.   This acid sulphate is a powerful reagent for the disassociation of certain kinds of iron ores, holding chromium, alumina, titanic acid, etc., since its excess of sulphuric acid is expelled only at a low red heat.]

Carbonate of Potassium.

Formula, $K^2CO^3 + 2H^2O$. Molecular weight, 174.

The aqueous solution of this salt should be limpid; neutralized by hydrochloric acid it should give no precipitate—nor any with the salts of barium (presence of sulphate), or the hydro-sulphide of ammonium (presence of alumina or of metallic salts). The solution, neutralized and evaporated to dryness in a capsule, should leave a residuum completely soluble in water.

Black Flux.

An intimate mixture of carbonate of potassium and finely divided charcoal is thus called. It is obtained by mixing one part of saltpetre and two parts of crude tartar (acid tartrate of potassium); this mixture is put into an iron vessel, and set fire to by an ignited piece of charcoal. After combustion, the residue is pulverized and kept in a dry place.

Chlorate of Potassium.

Formula, $KClO^3$. Molecular weight, 122.5.

The chlorate is tested in the same way as the nitrate, with regard to purity. It is obtained in commerce of great purity.

[Bichromate of Potassium.

Formula, $K^2O.2CrO^3$. Molecular weight, 295 2.

This salt is found in the trade in red tabular crystals, and is often contaminated with sulphate of potassium or sulphate of calcium. It is employed for the volumetrical analysis of iron, and the presence of the foreign sulphates is not objectionable

4

if the liquor be standardized by metallic iron, etc., and not from a given weight of the bichromate of potassium].

Permanganate of Potassium.

Formula, $K^2Mn^2O^8$.　Molecular weight, 316.

This salt is furnished in commerce in crystals of almost perfect purity.　It has long been designated " Mineral Chameleon."　As a general rule the larger and more regular the crystals are, the purer the salt.

Sulphocyanate of Potassium.

Formula, KCNS.　Molecular weight, 97.

The sulphocyanate of potassium, generally known by the name of sulphocyanuret of potassium, is only made use of for the qualitative determination of the per-salts of iron.　For use, dissolve one part of the salt in ten to fifteen parts of water.

[Potassium Ferricyanide.

This substance, often called *red prussiate of potash*, is employed for the volumetric determination of iron by the bichromate of potassium.　It should produce no precipitate or blue coloration with a ferric salt.　It is prepared by slowly passing chlorine through a cold solution of potassium ferrocyanide, until the liquor acquires a deep reddish-green color, and ceases to precipitate a salt of peroxide of iron].

B.—Salts of Sodium.

Chloride of Sodium.

Formula, NaCl.　Molecular weight, 58.5.

Chloride of sodium is used in quantitative determination of sulphur by dry assay, in order to reduce the intensity of action of saltpetre, with which it is mixed. It is indispensable, therefore, to have it completely free from sulphates. The testing for this is the same as in the case of nitrate of potassium, viz., by a barytic solution.

Sulphuret of Sodium.

Formula, $Na^2S + 9H^2O$. Molecular weight, 240.

This salt is deliquescent, and decomposes in the air; it should be kept in a well-stoppered bottle. It serves for volumetric determinations of zinc; for this purpose the crystallized sulphuret of commerce is of sufficient purity. A solution of sulphuret of sodium is prepared as follows: Make a lixivium of pure caustic soda, divide it into two equal parts, through one of which pass hydrosulphuric acid gas to saturation, then re-unite both parts, adding, if necessary, a little solution of caustic soda to remove completely the odor of the hydrosulphuric acid, and then filter to obtain a clear liquid.

Nitrate of Sodium.

Formula, $NaNO^3$. Molecular weight, 85.

This salt may be employed instead of the corresponding salt of potassium. It should be tested in the same manner. It is less energetic than nitre, and in commerce much more impure.

Sulphite of Sodium.

Formula, $Na^2SO^3 + 10H^2O$. Molecular weight, 306.

The aqueous solution of this reagent should be limpid; heated with sulphuric acid to the complete expulsion of the sulphurous anhydride, it ought not then to be affected by the hydrosulphide of ammonium.

[The solution of the resulting sulphate of sodium, heated with a solution of molybdate of ammonium in nitric acid, should not become yellow, as that color will indicate the presence of silica, arsenic, or phosphorus.

The commercial sulphide of sodium, in the solid form, sometimes contains carbonate of sodium, which, in nearly neutral solutions, may occasion precipitates. It is preferable to employ the sulphite of sodium with an excess of sulphurous anhydride, known under the old name of bisulphite of soda.]

[*Hyposulphite of Sodium.*

Formula, $Na^2S^2O^3$.   Molecular weight, 158.

This salt is found in the trade in dry, clear, and well-formed crystals, which should rapidly dissolve in water. The solution should not effervesce with acetic acid, and, when acidified, must, after a short time, become milky from the separation of sulphur.

   The hyposulphite of sodium is used for the precipitation of several metals, and for volumetric analysis. Like the sulphite of sodium, it possesses a deoxidizing action, that is to say it becomes oxidized itself.]

Carbonate of Sodium.

Formula, $Na^2CO^3 + 10H^2O$.   Molecular weight, 286.

This combination should comply with the same conditions as to purity as the carbonate of potassium. It is used like the latter for breaking up by fusion, mineral bodies insoluble in acids. Instead of using either of these salts separately, a mixture of the two is to be preferred as being more fusible than either alone. The mixture is composed of thirteen parts of carbonate of potassium, and ten parts of carbonate of sodium, dry, and finely powdered.

We shall designate this mixture as sodic carbonate of potassium.

Metaborate of Sodium (Glass of Borax).

Formula, $Na^2Bo'O^7$.   Molecular weight, 202.

This salt is employed as a flux, in dry assays. Commerce furnishes it sufficiently pure occasionally, but often containing silica, alumina, and iron in minute quantity.

Phosphate of Sodium.

Formula, $Na^2HPO^4 + 12H^2O$.   Molecular weight, 358.

This should dissolve in water without leaving any residue, and the solution ought not to be affected, even when warm, by ammonia.   For use, dissolve one part of the salt in ten parts of water.

[The phosphate of sodium of commerce is often efflorescent, from the presence of carbonate of sodium.   It should be purified by recrystallization.

The precipitates produced by solutions of the nitrates of silver and barium poured into that of phosphate of sodium, should, without effervescence, redissolve upon addition of dilute nitric acid.]

Acetate of Sodium.

Formula, $C^2NaH^3O^2 + 3H^2O$.   Molecular weight, 136.

The solution of this salt should be limpid, and remain so after the addition of oxalate, or hydrosulphide of ammonium.

Succinate of Sodium.

Formula, $C^4H^4Na^2O^4 + 6H^2O$.   Molecular weight, 270.

Succinate of sodium has the advantage over the corresponding salt of ammonium of being found ready prepared in commerce; nevertheless, the succinate of ammonium is to be preferred, because it does not leave any fixed residue. Tests for purity the same as for the acetate.

Nitroprusside of Sodium.

Formula, $FeNa^2(CN)^5NO + 2H^2O$.

This combination is only used for qualitative analysis. The crystallized salt may be had in commerce. It is not much employed, though its reactions are numerous and distinctive in some instances.

C.—SALTS OF AMMONIUM.

Chloride of Ammonium.

Formula, NH^4Cl. Molecular weight, 53.5.

Chloride of ammonium, heated on platinum foil should evaporate without residue. The solution should be colorless, and should give no precipitate with the hydrosulphide of ammonium. For use, employ five parts of water to one of the salt. The sublimed sal ammoniac of commerce is tolerably pure, but usually contains a little iron.

Hydrosulphide of Ammonium.

Formula, $(NH^4)HS$. Molecular weight, 51.

This is obtained by super-saturating ammonia with hydrosulphuric acid gas; it is generally purchased ready prepared. It must be preserved in well-stoppered bottles, and excluded from the light.

Molybdate of Ammonium.

Formula, $(NH^4)^2MoO^4$. Molecular weight, 196.

This reagent is employed in solution in nitric acid; one part of molybdic anhydride is dissolved in eight parts of ammonia and twenty parts of nitric acid. It is filtered, if it be necessary to obtain a clear solution.

[The mixing of the ammoniacal solution of molybdic acid with nitric acid, is attended with the production of heat, which, if not checked by pouring the acid by small quantities at a time, or by cooling the vessel, may cause the precipitation of a part of the molybdic acid. The temperature should not be allowed to go over 40° C. $= 104^{\circ}$ F.

The nitric solution should be allowed to stand for several days, until complete precipitation of any phospho-molybdate of ammonium, or of other combinations with silicic acid that may be present. The clear liquor is separated by decantation, or by filtration through clean asbestos, which has been previously washed with hydrochloric acid.]

Sesquicarbonate of Ammonium.

Formula, $(NH^4)^4C^3O^8 + 3H^2O$. Molecular weight, 290.

The test as to purity is the same as that for chloride of ammonium. For neutralizing acid fluids it is preferable to use this salt in the solid state; as a precipitant it should be in solution. One part of the salt to four of water should be used; one part of ammonia is added in order to saturate the excess of carbonic acid contained in the carbonate of commerce, which is the sesquicarbonate of ammonium represented in the formula.

Acetate of Ammonium.

Formula, $C^2H^3(NH^4)O^2$.   Molecular weight, 77.

The acetate of ammonium (the test for the purity of which is the same as that for the carbonate) may be employed in preference to the acetate of sodium, because it does not contain fixed matters, but its high price generally excludes it from industrial laboratories.

Oxalate of Ammonium.

Formula, $C^2(NH^4)^2O^4 + H^2O$.   Molecular weight, 142.

Test for purity, the same as that for the preceding salts.   The oxalate is employed to precipitate lime. One part should, for this purpose, be dissolved in twenty-four of water.

Neutral Succinate of Ammonium.

Formula, $C^4H^1(NH^4)^2O^4$.   Molecular weight, 152.

The crystallized succinate of ammonium furnished by commerce, is the acid succinate, $C^4H^5(NH^4)O^4$; molecular weight, 135.   The neutral succinate is prepared when required, by means of succinic acid directly, or the acid succinate dissolved in water may be saturated with ammonia until it presents neutral reaction on litmus paper.

VI.—SALTS OF THE ALKALINE EARTHS.

A.—Salts of Barium.

Chloride of Barium.

Formula, $BaCl^2 + 2H^2O$.   Molecular weight, 244.

This should be completely soluble in water, and should not contain any salt precipitable by hydrosulphide of ammonium. It is used in solution with ten times its weight of water.

Nitrate of Barium.

Formula, BaN^2O^6.   Molecular weight, 261.

Test for purity the same as that for the chloride. The solution is made with fifteen parts of water to one of the salt.

Carbonate of Barium.

Formula, $BaCO^3$.   Molecular weight, 197.

It is necessary to prepare the carbonate of barium in the laboratory, because that had in commerce, having been desiccated, will not answer. The preparation is as follows: Dissolve chloride of barium in a large quantity of warm water, heat the solution, and as soon as it commences to boil, pour in gradually a solution of carbonate of ammonium or of sodium until precipitation takes place, then let the fluid settle, protected from dust; decant the supernatant clear liquid, and repeat the washing by decantation with warm water until the supernatant liquid gives no precipitate with nitrate of silver. The carbonate ot barium is again suspended in as much water as will form a cream or pap, and preserved in that state in a stoppered bottle.

[It is very difficult to remove entirely the carbonate of sodium, by washing, from the carbonate of barium. It is preferable to employ the carbonate of ammonium for the precipitation. After a thorough washing, a few drops of hydrochloric acid may be added to the precipitate, thus forming a small proportion of chloride of barium, which, by standing for a few days, will decompose what may remain of alkaline carbonate. Another thorough washing will complete the operation.

The presence of an alkaline carbonate in the carbonate of barium is very objectionable, since it may occasion the precipitation of substances which are intended to remain in solution in the filtered liquor. On this account, several chemists prefer employing the finely powdered natural carbonate of barium, which, however, is open to the objection of being frequently mixed with metallic oxides.]

Acetate of Barium.

Formula, $C^4H^6BaO^4 + 3H^2O$. Molecular weight, 304.

This is very easily procured by saturating acetic acid with carbonate of barium. It is a reagent of very limited use.

B.—Salts of Calcium.

Chloride of Calcium.

Formula, $CaCl^2$. Molecular weight, 111.

Chloride of calcium is only employed in virtue of its strong deliquescence as an absorbent of water vapor, so it need not be pure. It can be easily prepared by completely saturating hydrochloric acid with white marble chips, and evaporating the solu tion in a porcelain dish until the mass becomes pasty, porous, and dry, or the heat may be raised so as to fuse the salt. It is reduced to fragments, and kept in a well-stoppered bottle. The merely dried salt is preferable to that which has been fused, since the latter always contains a little lime arising from decomposition of some of the chloride by the combined action of heat and water; besides, the porosity of the dried chloride gives it an absorbent power much superior to that of the fused chloride.

Carbonate of Calcium.

Formula, $CaCO^3$. Molecular weight, 100.

This reagent is employed for the quantitative determination of iron by dry assay. We shall treat of it under that head.

C.—SALT OF MAGNESIUM.

Sulphate of Magnesium.

Formula, $MgSO^4 + 7H^2O$. Molecular weight, 246.

For use it is indispensable that this solution, mixed with one of chloride of ammonium, be not affected either by ammonia or by its oxalate, nor by hydrosulphide of ammonium, even after an hour's repose. This salt serves to precipitate phosphoric acid. For this purpose prepare a solution of one part of crystallized sulphate of magnesium and one part of chloride of ammonium in eight parts of water; add four parts of ammonia, let it rest for some days, and then filter.

VII.—METALLIC SALTS.

Nitrate of Silver.

Formula, AgNO^3. Molecular weight, 170.

The commercial salt sold for photographers' use is very pure: one part should be dissolved in twenty parts of water, and the solution preserved from the fumes of the laboratory and from light.

It has been known for a considerable time that the blackening of nitrate of silver, or its solutions, by light depends upon the presence of organic matter, including in that the dust ordinarily floating in the air. If the solution be made with distilled water, free from organic matter, and the nitrate of silver solution bottle be kept in a tin plate or wood case, and washed, rather than wiped with a cloth, externally, there will be but little of that blackening so generally and disagreeably found about the mouth, etc.

Chloride of Nickel.

Formula, NiCl^2 + 9H^{2}O. Molecular weight, 291.7.

The crystallized chloride of nickel of commerce is pure enough for the purpose for which this reagent is intended. It is used in solution in water.

Chloride of Copper.

Formula, CuCl^2. Molecular weight, 134.5.

This reagent is employed in solution prepared either by dissolving in water the crystallized salt, or by the direct solution of copper in aqua regia containing an excess of hydrochloric acid, evaporating the solution to dryness, and treating the dry mass with water.

Sulphate of Copper.

Formula, $CuSO_4 + 5H_2O$.   Molecular weight, 249.5.

The salt of commerce purified by recrystallization will answer.

[Acetate of Copper.

Formula, $C_4H_6CuO_4 = Cu2(C_2H_3O_2)$.   Molecular weight, 181.5.

This salt is prepared by dissolving verdigris in hot acetic acid, and leaving the filtered solution to cool.   It forms dark-green crystals, soluble in water and alcohol.

It is employed for the purification of precipitates of sulphate of barium, which has a great tendency to carry with it nitrates and chlorides present in the liquors.

The solution is prepared by dissolving the crystallized acetate of copper in hot water, with a little acetic acid and two drops of sulphuric acid, and then adding a few drops of chloride of barium, enough to give a slight barium reaction.   Boil a short time and filter.   By cooling, crystals are deposited, and the supernatant saturated solution is employed.]

Nitrate of Lead.

Formula, PbN_2O_6.   Molecular weight, 331.

Nitrate of lead can be had in commerce of purity sufficing for our purpose.

Acetate of Lead.

Formula, $C_4PbH_6O_4 + 3H_2O$.   Molecular weight, 379.

The remarks as to nitrate of lead apply here.

Chloride of Mercury.

Formula, $HgCl_2$.   Molecular weight, 271.

The sublimed chloride of mercury known as "Corrosive Sublimate" is used.   This salt is a very violent poison.

5

Chloride of Tin.

Formula, $Sn^2Cl^4 + 4H^2O$.   Molecular weight, 450.

The commercial crystallized chloride of tin is used. It is dissolved in water acidulated with hydrochloric acid. A solution of this chloride may also be prepared by dissolving grain tin in hydrochloric acid, with heat.

Protosulphate of Manganese.

Formula, $Mn^2S^2O^8 + 8H^2O$.   Molecular weight, 446.

The solution of this reagent in water, acidulated with sulphuric acid, should be clear and colorless, and should not decolorate permanganate of potassium.

Perchloride of Iron.

Formula, Fe^2Cl^6.   Molecular weight, 325.

To obtain a solution of this reagent we should treat metallic iron with hydrochloric acid, diluted with its own volume of water, taking care that the iron should be in excess: filter and pass a current of chlorine into the liquid until it is saturated; then boil in order to get rid of the excess of chlorine. Or hydrated peroxide of iron precipitated by ammonia may be dissolved in hydrochloric acid.

[Instead of the troublesome preparation of chlorine, a few drops of bromine may be employed, and the solution boiled. The peroxidization of the iron may also be effected with chlorate of potassium and sufficient hydrochloric acid to entirely decompose the chlorate, and the solution evaporated to dryness.]

Sulphate of Iron and Ammonium.

Formula, $Fe(NH^4)^2S^2O^8 + 6H^2O$. Molecular weight, 392.

This salt is seldom to be had pure in commerce; it is best to prepare it when required. Take two equal portions of dilute sulphuric acid, neutralize one with ammonia or the carbonate of ammonium, then add some drops of the acid until it reddens litmus paper; heat the other portion with iron free from manganese, until gas has completely ceased to come off, leaving the metal in excess. Filter both liquors while warm, then pour them together and leave them to crystallize. Decant the mother liquor, and then wash the crystals in a little cold water, and let them dry in the air between sheets of filtering paper. The solution of this double sulphate should be limpid, and should not turn red immediately on the addition of sulphocyanate of potassium.

[This double salt, when pure, contains exactly one-seventh of its weight of metallic iron]

VIII.—TEST PAPERS.

Blue Litmus Paper.

Digest one part of commercial litmus in six parts of water. Filter the deep blue liquid and divide it into two equal parts; into one, drop carefully some very dilute sulphuric acid, until the blue color begins to show a tinge of red; unite the two parts in a sufficiently large capsule or deep plate, and dip in some sheets of unsized paper; dry the sheets where no acid vapor can reach them; cut them into small strips, and preserve them in a wide-mouthed stoppered bottle.

Red Litmus Paper.

This is prepared by reddening blue litmus solution used for the blue paper, by adding some drops of very dilute sulphuric acid and soaking the strips of paper in that liquid, and then drying.

Salt of Lead Paper.

This test paper is only used in the volumetric analysis of zinc by sulphide of sodium. The paper employed is glazed, cream-laid note-paper, or that used for visiting cards.

REMARK.—Before concluding this Part, a general remark may be made on the most advantageous manner of employing the reagents just described, especially those which are used in solution. It is well to know the standard or value of dissolved matter in all such solutions, and our advice is to inscribe on the different test-bottles labels similar to that we give as an illustration with respect to chloride of barium.

(One) Cent. cubic (contains):—

0^{gr}.10 chloride barium cryst^d ($BaCl^2 + 2H^2O$)
(corresponding to) 0^{gr}.0131 of sulphur.
"   " 0^{gr}.0328 of SO^3.

The words in brackets may be omitted as being implied.

In this way, and with a little practice, we avoid employing too large proportions of reagents, an excess of which is often far more injurious in the subsequent steps of analysis than is to be inferred from the mere waste of material.

Test bottles and their labels are a source of trouble in all laboratories, but the amount may be lessened by a few judicious arrangements often not known or not attended to.

The glass-stoppered bottles themselves require to be well chosen —the necks not too narrow, and with sufficiently wide, thin, and well-turned lips, to enable the solutions to be dropped or poured in a fine stream at pleasure; steady dropping is impossible with thick, lumpy, ill-rounded lips. The stoppers, though ground finely, so as to be air-tight, should not have an insufficient taper (the fault of all French bottles), for if so they get set fast, and the bottle often must be destroyed to extract them. No stopper, if properly tapered and well ground, need be more than one diameter in length; excessive length is the great objection to French stoppers. The flat-topped (round or octagonal) glass stoppers so common on German bottles are, on the whole, the best for solutions. They admit of more force to extract them when set fast, than the common vertical and compressed form of stopper. They protect the lip a little from dust, but they are less easy to wash or wipe quite clean, and when chipped anywhere at the edge are dangerous to the hands. Bottles holding rather less than a pint, and of not too thick glass, are the pleasantest in use, with stone bottles of larger size. In testing, glass stoppers taken out should not be laid upon the table, but upon a plate of glass or porcelain to preserve their purity. Glass cases for test bottles are always more or less troublesome, but their use is, after all, desirable, especially in laboratories established in the dusty precincts of iron works or manufactories, to prevent lodgment on the lips, etc., of the bottles. The fronts of glass cases for this purpose are best made to slide in grooves, or balanced to throw up towards the ceiling; when hinged as doors they are liable to come in contact with vessels or apparatus.

The labelling of test bottles, and of indeed all bottles in the laboratory, is a matter also in which a little forecast avoids much

5*

occasional trouble and wasted time in rediscovering the contents
of bottles with labels fallen off or obliterated by acid fumes, etc.
For large store bottles, the labels are best etched or written with
a scratch diamond direct upon the glass; and for the test acid
bottles this may be employed, but unless the marking be very
strong it is difficult to see, except in certain lights. For all other
test solutions, paper labels are best. These should be written
with good black common ink, mixed with some ground Indian
ink, which as regards acid fumes is practically indelible; or, so far
as they can be procured, the printed labels for reagents now sold
may be used. Gum or dextrine, as commonly used for causing
the paper to adhere to the glass, is objectionable. Damp or
laboratory vapors soon decompose these, so that the labels drop
off. A species of isinglass glue is sold very commonly in France,
and also to be had in London, known as "*colle forte a froid*,"
which is the best substance the editor has found for this use.
Varnishing the outer surface of the label has been occasionally
practised, but is objectionable, as in wiping the bottles the
varnished surface is scratched, and dirt then collects upon and
defaces the labels. If varnished, Canada balsam in thin solution
is perhaps the best. The so-called enamel-labelled reagent bottles
sold by chemical apparatus dealers are objectionable; the labels
are too large, and obscure the state of the interior of the bottle,
so that accurate pouring is less easy, from not seeing the whole
surface of the solution, etc.

PART II.

APPARATUS AND OPERATIONS.

A few suggestions may here be offered with advantage as to the general arrangement of an analytical laboratory, more especially in ironworks and other manufactories. A laboratory is decidedly best placed one story, or at least some feet, above the ground level: directly on the ground, the damp is always objectionable, and the floor cold and rheumatic for the feet and legs. The whole floor is best on brick arches, but the place about the analytical tables, where the operator chiefly remains, should be boarded, and that covered in winter with a strip of cocoa-nut matting. Few people realize how much the activity of the mind, upon which rapid and sure chemical progress is so much dependent, is interfered with by constant, though perhaps unheeded, bodily discomfort. The main light except in hot climates, is best derived from a large top light, which *may* be made to open upon exceptional occasions, but the ventilation should be otherwise arranged, for dust, especially in manufactories, should be carefully excluded.

There should be, however, at least one good window or side light, because, amongst other reasons, test tubes, etc., must often be held between the eye and the light to judge of color, etc. In laboratories (as for students) where there are a large number of operators, it is usual and necessary to place the working tables about the middle of the room, where the light is not so good ; but when there is only an analyst and a few assistants, we much prefer the working tables to be close against the side light. The windows, if more than one, should be all at one side ; cross lights are bad.

The north aspect, if obtainable, is the best for the windows: a glaring sunshine coming into the room, often succeeded by gloom in our English climate, interferes with perfect vision.

A room of about 30 feet by 25 feet, and 15 or 16 feet high, is large enough for much work. Three smaller rooms are best attached to it—viz., a store-room, for apparatus and reagents; a balance room, and a writing room, which shall also contain some few volumes of standard works of reference on chemistry, etc. These rooms are best at the entrance end of the laboratory proper, and the approach to the whole should be through a small corridor and ante-chamber, where coats and hats can be laid aside. This, especially in factories, should have double doors to exclude dust and draughts as far as possible. Fuel, fire-wood, coarse laboratory stoves, carboys of acid, distilled water, etc., should have one of the arched rooms beneath appropriated to them.

Another should be devoted to the preparation of certain reagents to be hereafter referred to; and from a third the fresh air for ventilation of the laboratory should be taken off. The external air should be brought in, sifted from dust, through a large sheet of copper wire gauze (brushed occasionally), and thence conducted up through flues of about $15'' \times 7''$ in the arched floor and walls of the laboratory, to six or more apertures, discharging through regulating ventilators into the room at about midheight between floor and ceiling, or, if desired, still higher.

In rigorous climates means should be provided for warming the supply of fresh air in the lower apartment by a coil of hot water or steam pipes.

The entrance to the laboratory being at the west end, and the northern windows thus as we enter on the left hand, the further end will be the best place for the furnaces, the floor for some six feet wide in front of these being tiled or bricked. Prominent amongst the furnaces should be the flat sand-bath, which also gives the means ordinarily of warming the room. An iron hood should extend over all the furnaces, and for the space of the sand-bath and the iron plates around it, should be formed into a glazed-in chamber; the wall here being lined with glazed white tiles, and a movable shelf or two of thick plate glass (nine inches wide) projecting from the face, and resting on pottery or glass brackets. These are very useful for laying aside slowly-precipitating solutions, etc., requiring some hours' repose in a warm place.

Also a good and capacious drying-closet or oven beneath the

sand-bath is very useful. The glazed iron fronts of sand-bath chambers such as these, are commonly hung like window sashes, with balance weights to throw up and down, but this plan never works smoothly or well. It is much better that the front consist of two or three panes of thinnish sheet glass loosely secured by pins into rectangular wrought-iron frames, of a light sort, hinged at top, and balanced so as to throw up towards the ceiling, moving radially round the top edge. Being balanced, any one of these remains up until pulled down and latched by a catch at the bottom edge. The writer has experienced the comfort of this arrangement in the place of the usual ones, in which the glass gets continually broken, by jamming of the iron sashes, by expansion, etc., or by unequal heating of the glass plates.

About 6 or 7 feet by 3 feet will be occupied by this sand-bath furnace ; and where practicable it is best that the fire-place, or small stoke-hole of it, should open (through the wall of the laboratory) in an external shed, so that the fire can be lighted and kept up by an attendant from the outside, and so the dust, etc., of a fire fed from within the room be avoided.

The draught from the sand-bath fire may be conducted through a wrought-iron tube for some 15 feet or so in height, the tube being placed within a pretty capacious brick flue formed in the wall. This flue is kept sufficiently heated thus, so that the air from the laboratory, close to the ceiling level, being conducted into it, by means of six or more ventilating apertures in the walls (intermediate to those for introduction of fresh air), a good out-draught shall be secured, and so ventilation kept up without opening doors or windows, the draughts from which derange the gas flames on the tables, and introduce dust, etc.

The northeast corner beneath the hood should be occupied by a stout wrought-iron plate or table about 5 or 6 feet long by 3 feet wide, at the same level as the sand-bath plates. This is very useful for combustions, for small furnace operations, etc., the wall being here provided with several apertures leading into one or more flues. Next to it, southward, stands the sand-bath, and further south, beyond it, the other furnaces. The iron table is best with a glazed front to throw up like that of the sand-bath, so that protracted gas evaporations, etc., can be conducted within it without annoyance, hence coal gas should be laid on here as well as to the tables.

The other furnaces usually needed are—a muffle furnace, both for assay by cupellation, and for scorifying ores, etc. ; one, or perhaps more, assay melting furnaces, for which no better model can be had than that of those habitually in use in Cornwall : and a good wind melting furnace, capable of melting wrought iron, and large enough to receive a crucible of 6 inches diameter. A chimney of 40 feet high will give sufficient draught: but in factories, when possible, the laboratory furnace flues are led away to some neighboring lofty stack.

At about the centre of the south side, a pretty capacious aperture through the wall should be occupied by a gas reaction chamber. The bottom or table of this should be about $3\frac{1}{2}$ feet from the floor, light admitted by the exterior surface of the chamber being close glazed, and the chamber cut off from the room internally, by a glazed, well-fitted sliding door. None of the woodwork of this or of the laboratory generally should be painted, but only varnished. A ventilating tube should lead out of this gas chamber to the out-draught ventilating flue, and an in-draught tube from the fresh-air flues.

In a well-provided laboratory a store of two at least, if not three, of the great gaseous reactives, viz., sulphuretted hydrogen, chlorine, and carbonic anhydride, should be maintained in the lower room already referred to. These gases are generated in large apparatus on the principle of Döbereiner's, already described in the text, in which a few cubic feet of gas can always be kept under the pressure of two or three feet of liquid. From these the gases should be conducted up in distinctively colored leaden, glass, or gutta-percha tubes, to the gas chamber, and there provided each with a stop-cock, and each with half-a-dozen small India-rubber nozzle tubes, to which the glass tubes for immersion in the solution to be acted on can be attached. This storage of reactive gases saves a vast amount of time, and with the thus-constructed gas chamber saves stenches in the laboratory and injury to health.

A supply of water, ample and unfailing, must be provided, and this should command a height of a few feet above the level of the ceiling, in order to admit of what may be thought a luxury, but which is in fact one of the most valuable adjuncts that can exist in a laboratory, viz., an instrument of the nature of the Tromba, or water blast, which affords at all times a powerful blast of atmospheric air—either for small furnaces, or gas or other large blow-pipes, and of exhaustions analogous to that of the Sprengel air-pump. This, as constructed at the Physical Laboratory of the College de France, at Paris, by M. Jamin, consists of a cylindrical copper vessel of about 24 inches long by 10 inches diameter, with its axis horizontal, placed above the laboratory ceiling (in the roof or apartment above); into this, water from the supply is introduced through a pipe ($1\frac{1}{4}$ inch diameter), with a stopcock for regulation; from the lower side of the cylinder, three or four pipes, each about $\frac{3}{4}$ inch diameter, descend to the level of the ground below the laboratory. Each of these has two smaller air-pipes soldered into it, issuing from two horizontal tubes, a a, close to the cylinder above the water level therein, and branched into the descending water-pipes close beneath the

Fig. A.

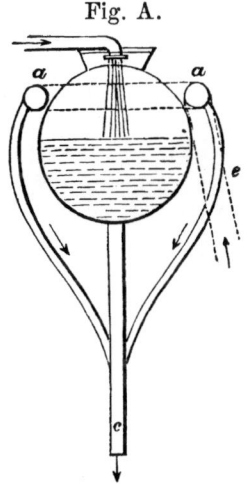

cylinder, as shown in Fig. A. By these air is constantly drawn in through a a (these tubes join into one) and passed down with the descending current of water in the tubes c.

At the lower end, the tubes c pass into a copper cylindrical vessel with open bottom, inverted into a small cistern of water, the two being much like the arrangement of a gasometer (Fig. B). In this the drawn-down air under the pressure due to the

descending column, and to the depth of water in the bottom
cistern, is separated from the water, the latter running off to
waste.

Fig. B.

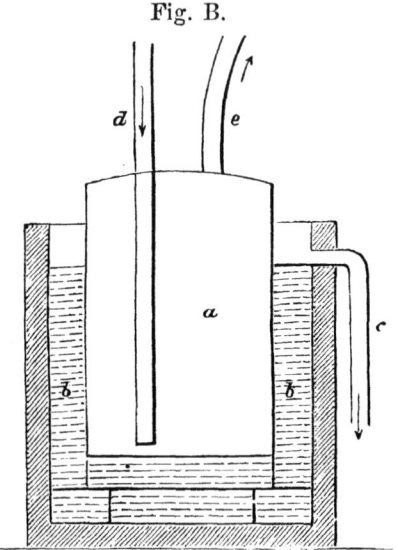

a. The copper and water separating vessel, open-bottomed.
b. The cistern containing it.
c. The water overflow, or waste.
d. The descending water and air-pipe.
e. The air blast pipe ascending to the laboratory.

A pipe (e, Fig. B) is led from the top of the copper air vessel
at bottom up to the laboratory ; and it is only necessary to open
the stopcock thereon to have at once a steady, powerful blast,
which may be regulated at will. The possession of such a blast
enables work to be done with gas burners or gas furnaces that
would scarcely be credited by those who have only employed
that of table bellows. With the form of bent gas blowpipe,
much used in Birmingham for brazing and soldering, etc., which
is held in the hand and supplied with air and gas by flexible
tubes, a powerful blowpipe brush of flame can be thrown in any
direction ; and, fusions with alkalies, etc., of minerals, may be
effected in a few minutes in platinum crucibles, with ease and
cleanliness. For glass-blowing also, this beautifully steady blast,

which leaves the bodily movements of the operator free from working bellows, etc., is invaluable.

Conversely, means being provided at the top cylinder (Fig. A) for putting the air apertures, *a a*, all uniting in the tube *e*, into connection with a pipe leading up for some part of the laboratory, it is obvious that exhaustion or aspiration is effected, with any degree of power up to that of the descending head of water. The main aspirator pipe in the laboratory is furnished with several branches and stopcocks, which are brought to the tables reserved for filtration, and give the means, by the method of Bunsen, of greatly expediting these tedious operations. At the same time, the degree of exhaustion at its full amount is sufficient to maintain a moderate partial vacuum beneath large bell glasses, with sulphuric acid dishes included, for desiccation or evaporation at atmospheric temperature.

In a laboratory situated in a factory, a constant supply of steam from some adjacent steam-boiler is of great value, and gives the means of heating a drying-closet for filters, etc., which may then be all of glass and wood, or, if desired, of warming the laboratory itself. Thus, with plenty of water, coal gas, for heat and light, steam, air-blast, vacuum, and reactive gases—all to be had by turning a cock—time becomes economized to the highest degree.

With steam, too, boiling water for washing purposes can be had, as well as cold water for the washing sink. This last should be capacious, but of strong glazed white pottery, and with a lead-covered and slightly-inclined table adjacent, to lay soiled and cleansed glasses, etc., upon. The main sink is best placed at the north side of the entrance door, in entrance end of the laboratory, the wall at the other side of the door being occupied by presses for reagents, preparations, etc. The table blowpipe, and apparatus for gas heating, may occupy the south side to the east of the gas reaction chamber ; and at the other side of that an air-pump, which is best on Deleuil's construction, where furnace or other gases are to be examined. This is also the place for the pneumatic troughs.

The general illumination of the room artificially, is best effected by Argand shaded burners, suspended over head from the middle of the ceiling. For night-work, a gas-light is also needed at

6

each working table; but chemical operations (like those of the painter) are badly conducted unless by day-light.

The reactive bottles in constant request are in student laboratories generally placed on shelves elevated from the centre of the table; but that is a bad arrangement for analytical work, and we prefer that they should rest on narrow shelves occupying the piers between the northern windows, and so be close to the operator's left, leaving his tables quite free. Three northern windows afford spaces for three working tables, and four piers for shelves for reagents. There should be a small white pottery sink at the south end of each of these tables, close at the operator's right hand, his reagents being on his left hand, and his table and work before him. Thus the three tables afford space for an analyst and two assistants. In a very extensive analytical industrial laboratory, perhaps a distinct room for volumetric assays might be advisable. We have, however, sketched the arrangements for a first-class analytical laboratory for general industrial uses, and to which there may be many different degrees of approach.

The arrangements may also need to be sometimes, though rarely, somewhat modified to meet the special requirements of some particular trades—as, for example, those dealing with oils or soaps, etc. But we are here to keep in view principally the analytical laboratory for metallurgic uses, and especially those related to iron manufacture.

Upon one thing capitalists and proprietors of large ironworks may rely—namely, that make-shift laboratories, like make-shift tools and machinery generally, are the most expensive in the end.

If it be worth while to take aid from chemical science at all, it is cheapest and best done by providing a competent analyst, with the best and most convenient instruments and apparatus that money can procure. It is evident, however, in most instances, in Great Britain at least, that this view is as yet not acted on.

[If metallic hoods are desired, we prefer, from experience, those of sheet zinc to those of iron, provided, however, that the flame of a muffle furnace is not made to act upon the zinc.

Wooden hoods plastered one inch thick, inside and outside, with plaster of Paris, are quite satisfactory. The plaster is held to the wood by studding the latter with a sufficient quantity of iron nails. The coat, when dry, is rendered impervious by smearing it several times with boiled linseed-oil. Some persons use simply a white-washing of slaked lime in whey. Such hoods should not be too near the fire.

In a similar manner, the tables for combustions and small furnace operations, may be of wood with a cover of tightly-fitting bricks, well cemented.]

Balances.

An analytical balance should be sensible to less than a milligramme. It ought to be well preserved from the fumes of the laboratory and from dust; a special place or room should be set apart for it, and it should be inclosed in a glass case. In order to

Fig. 5.

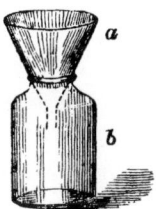

avoid the oxidizing action of moist air, an absorbent substance ought to be placed inside the case; the arrangement which we consider desirable is repre-sented in Fig. 5. A funnel *a*, containing pieces of chloride of calcium, rests in the neck of a bottle, *b ;* as the chloride attracts the atmospheric water, the concentrated solution which is formed flows into the bottle, still leaving the surface of the salt exposed. After some time we can take out the liquid, re-obtain the dry chloride of calcium by evaporation, and replace it in the funnel.

A shallow, light wooden tray, partly filled with about half a pound of dry, unslaked lime in small fragments, is, in the editor's judgment, greatly to be preferred to the chlor-calcium apparatus of the text. If the latter be once upset and spilled in the balance case, the balance itself becomes soon corroded or useless.

Dry lime is sufficiently hygroscopic, and if spilled does no harm. A tray equal in length to the glass case, and not more than 2 inches wide by $1\frac{1}{4}$ inches deep, and supported on two little wood slips close up to the top of the glass case, and behind the beam, has been adopted with convenience by the writer.

It is most important to avoid rough or careless handling of the balance, to remember never to place weights or any substance whatever on the pan or plate, until the apparatus be properly adjusted. Balances which have an arrangement to support the beam off the knife edges, and relieve it of the plates or pans, when not in use, are very desirable; objects can then be left upon them, and oscillations which interfere with weighing are thus avoided. There are balances in which the beam, or at least one of its arms, has a decimal division, intended to mark the position of the small fixed weights called "riders," which act on the principle of the steelyard. The use of the "rider" is much more convenient and exact, than that of the weights of one, two, and five milligrammes.

Weights.

It is important to ascertain the accuracy of the weights provided. If the system of weights be exact, each single weight tried, with a just and sensitive balance, should balance exactly its value in subdivisions or smaller weights.

Weighing.

We should never weigh an object while warm; in that case a current of ascending air is produced which will make the object appear lighter than it really is.

We should also avoid weighing objects whose temperature is much lower than that of the air of the balance-room; in this case a current of descending air is produced, and also vapor of water may condense on the object, and these united causes will increase the apparent weight.

In order that bodies may attain the suitable temperature without absorbing moisture from the air, we should place them in a drying bell-glass. This apparatus is simply a vessel containing some chloride of calcium or sulphuric acid intended to absorb the moisture.

To weigh a precipitate that cannot be calcined, we must collect it on a previously weighed filter, and dry it at a temperature of 100° Cent. Choose a filter of fit size and well washed, place it in a stove heated to 100°, and with it two watch-glasses of the same diameter which fit exactly over each other. When the drying process is completed, we place the filter between the two glasses, fastening them together with a slender brass clip, which has also been placed in the stove. The whole is left to cool in the desiccating bell-glass, and then weighed. When the filtering has been accomplished and the precipitate well washed, heat it to 100°, first in the funnel and then on one of the watch-glasses. After complete desiccation, apply the second glass, and fasten both

6*

together with a clip. When all has been again cooled in the drying bell, weigh again, the difference in weight is the weight of the precipitate.

Weighing precipitates upon the filter should never be done, unless through absolute necessity. Whenever practicable, it is best to redissolve a precipitate so circumstanced again, and get it into some stable form that can admit of being desiccated, or even calcined, in a porcelain or platinum capsule, and weighed therein. Where weighing with the filter cannot be avoided, a *very* thin short test tube of about eight-tenths of an inch diameter, with a very light, well-fitting cork, is a lighter and better guard from hygroscopic change than double watch-glasses. The filter, loosely folded, is easily pushed in and taken out.

Most precipitates are weighed after calcination. This operation demands certain precautions, some general, and some depending on the nature of the combination. A precipitate is never calcined until dried. The calcination is generally performed in a muffle; the matter to be calcined is first placed in front of the muffle until the filter be carbonised, then it is placed in the muffle and heated to redness. Care should be taken to raise the temperature gradually, if the precipitate be such as will give off gas under the influence of heat, such as the oxalate of calcium, and the biphosphate of ammonium and magnesium. When the precipitate is easily reduced (sulphate of lead, carbonate of zinc), it is necessary to separate the matter from the filter as much as possible, and to calcine each separately.

Calcination is generally performed in small platinum or porcelain capsules; the latter are only used when the nature of the precipitate will not permit it to be heated in contact with platinum.

Having weighed the calcined precipitate, we must not forget to abstract from the weight found—that furnished by the ashes of the filter, determined by previous investigations. We may allow that filters of the same size, made of the same paper, will always leave the same weight of ashes when care has been taken to wash them with dilute hydrochloric acid; without this last precaution the paper produces a larger quantity of ashes after the filtration of an alkaline or neutral fluid than after an acid one. It is important to provide a number of filters, cut by a pattern, of three or four different sizes, to steep them for ten minutes in pure hydrochloric acid, diluted with double its volume of water, then to replace this liquid with pure water as often as necessary—that is, until every trace of the acid has disappeared, which is ascertained by nitrate of silver. After the desiccation of the filters, we should burn a certain number of each size, and weighing the ashes, ascertain the quantity of fixed matter contained in each, and keep a note of it, as a constant.

Among all the different ways of weighing there is one which we must describe. It is used for ascertaining quantities with great accuracy. Take a tube from three to five millimetres in interior diameter, and from 15 to 20 centimetres long. Close this at one end, then weigh it, fill it nearly with the substance to be analyzed, and weigh again. We know very closely the quantity of substance contained in the tube by substracting the difference. We can then pour into the vessels prepared to receive it small portions of the substance in the tube, taking care to weigh the tube every time, so that by a series of substractions we ascertain the successive quantities introduced into each vessel. This method is also employed when the substance has to be introduced dry into an apparatus with a narrow mouth; for instance, into a little phial, or tube with a ball. In this case we should introduce into the weighing-tube only the quantity required.

If we are obliged to make several experiments for the complete analysis of a substance, we may reduce the number of weighings by dissolving a large quantity at once (10 grammes, for instance), diluting the solution thus obtained (having separated the insoluble residuum by filtration) with a known volume (500 or 1000 cubic cent.), and performing the operations with fractions (*e.g.* 50 or 100 c. c.) of the liquid measured off.

The method of "double weighing," so important in cases demanding scrupulous exactness, is not referred to by the authors, nor indeed many other matters—as to the construction and manipulation of analytical balances—of importance to be known. For these the works on Analysis of Rose and Fresenius, Watt's "Dictionary of Chemistry" (article, Balance), and the "Traité de Physique" of Jamin, should be studied.

The matter to be weighed is best always placed in the pan to the left of the operator, and the weights in that next his right hand. In making note of the weight ascertained, always do so before removing the weights from the pan: trust nothing to memory.

A good analytical laboratory requires at least two sizes of fine or exact balances (none are better than those made by Oertling, of London), one capable of weighing up to 500 grammes, or even a kilogramme, the other to about 50 grammes; with two larger beams (or German table weigh-bridges) for larger quantities needing little exactness. British grain weights are really just as convenient as grammes, but the whim of fashion with chemists now is for French weights and measures, the use of which sometimes simply is to produce an immense amount of trouble in comparing late with older works, etc.

Apparatus for the measurement of Liquids.

We should not neglect to test such apparatus, which are often constructed carelessly. There is occasion to distinguish gauged apparatus, which only serve to measure one certain volume or its subdivisions of liquid, and simply graduated apparatus, which will measure any volume whatsoever. The only gauged apparatus required in an industrial laboratory are flasks and tubes; they may be gauged in two ways—either by filling or by emptying; in the first case, they should contain, when filled to the mark, a volume of liquid equal to that indicated; in the second case, they should produce, by emptying the liquid which has been poured in up to the mark, the volume indicated there; the difference is the amount of liquid adhering to the sides. There are vessels with two marks, one lower, corresponding to gauging by filling; the other higher, corresponding to gauging by emptying. The apparatus with the two marks are certainly to be preferred.

The graduated apparatus employed are jars, cylinders, or tubes (usually called burettes). The most simple and convenient are *burrettes à pince*—that is to say, graduated tubes with a narrow neck at bottom and an India-rubber short tube, capable of being closed instantly, by a spring clip of wire.

They should always be graduated by emptying. The control of the graduation of all these vessels is

by weight. We must determine the weight of the
quantity of water at 16° Cent., which the space
between two marks contains, and that of the quantity of water which will exactly fill the apparatus.
Knowing this weight, we can easily tell the weight
of the volume, remembering that a litre of distilled
water at 16° Cent. weighs 999 grammes.

The advance in the manufacture of chemical apparatus of late
years is such that graduated or gauged vessels, and measures of
glass, can always be bought much better than they can be made
by the chemist. From the very fact, however, that these are
manufactured wholesale, they are not always constantly reliable
as to correctness; and hence those to be employed constantly and
for important purposes should be, in the first instance, carefully
verified by weighings of their assumed contents, and the constants
of error, if any, found and registered.

Filtering.

This is one of the most frequent operations; it is
often long and wearisome, and endeavors to diminish
its duration have, therefore, been attempted. The
rapidity with which a liquid filters depends, all
circumstances being the same, on the difference of
pressure exerted on each side of the filtering paper.
It is evident, then, that filtration can be hastened or
retarded by increasing or diminishing this difference
—i. e., by acting upon either side of the filter. It
would be difficult to increase the pressure on the
upper side, but it can easily be diminished on the
lower side. For this purpose the funnel should be
-attached to a matrass or bottle by means of a caoutchouc cork, furnished also with a bent glass tube.
This tube, with the aid of an India-rubber tube,

connects the matrass with an air-pump. If the pressure be more than very slight, the filtering paper will not sustain it; in that case it should be supported by placing in the funnel a cone made of a thin sheet

Fig. 6.

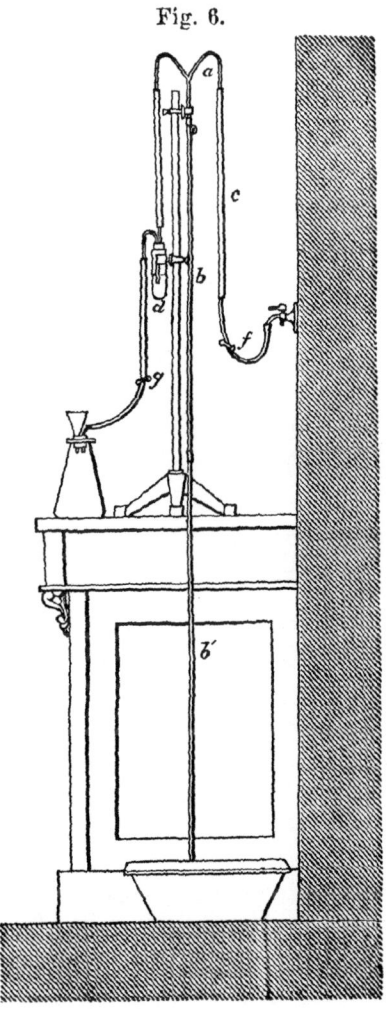

of platinum, twisted into a cornet or cone. Paper will stand the pressure of a metre of water, provided the orifice of the funnel be narrow; with a pressure of two metres it is prudent to employ a double filter.

Various forms of air-pump have been proposed, among which Bunsen's pneumatic pump, with water current (*Wasser luftpumpe*), should be placed in the first rank. One of us published in the *Berichte der deutschen Chemischen Gesellschaft*, 1870, p. 286, a description of an apparatus, the construction of which is on the same principle, but being simpler and of much smaller dimensions, would be found far easier to establish. The force of this apparatus is sufficient for the greater number of cases occurring in industrial laboratories. It consists principally of a tube of glass, *a*, in the form of a Y (Fig. 6), the lower part being in communication with the tube, *b*, *b'*. One of the side branches communicates with a reservoir or conduit of water, and the other with the vessel intended to receive the filtered liquid. On the tube uniting the apparatus to the vessel holding the funnel is a small flask *d* (*flacon laveur*), which regulates the working of the machine. The apparatus referred to on page 59 renders any other needless.

Drying or Desiccating Stoves.

Various kinds of stoves are used for desiccation according to the temperature required; some have a double bottom and sides intended to receive liquids which boil at various temperatures. In general, water is used, and in this case the temperature of the stove never exceeds 100° Cent.; others have single walls, and are called air-stoves, and heated directly. With the latter we can obtain a variety of temperatures more or less elevated. A stove with a double wall consists of a rectangular box of red copper, furnished with double walls on five of its surfaces; the front surface has a door with a small opening at the bottom for the circulation of the air; the upper surface has two tubular openings, one penetrating both walls and intended to hold a thermometer; the other communicating only with the outer chamber for the purpose of pouring in the liquid to be heated between the walls.

The form of the air-stove is precisely the same, but all its sides are formed of a single wall, and its upper surface is furnished with only one central opening intended to hold a thermometer.

These stoves on a small scale are usually made of copper. The steam drying-stove of glass referred to on page 61 is much more desirable.

Sand-Bath.

The sand-bath is formed with an iron plate or dish heated from beneath by a small furnace or fire-place, and covered above with a stratum of sand. Coarse quartz sand is preferable to fine bank or sea sand, as being a better conductor of heat, and because the latter adheres easily to the bottom of vessels, and in moving them some grains might fall from one vessel into another. The sand-bath should be covered by a glass case, divided, if possible, into two compartments, each communicating by a flue with a

7

good draught. If the arrangement of the place permit, the stoking-door of the fire-place should be outside, and in a room adjoining the laboratory, to avoid the dust produced by putting in coal. See page 56.

Water-Bath.

Among the different arrangements proposed for this apparatus, we select the one represented in Fig. 7 as the most convenient; *a* is a cylindrical metal vessel in which glass vessels of different sizes, as required, can be placed. A small apparatus, *b*, is attached to the lower part; this consists of two concentric glass tubes, the exterior fixed and the interior movable, in a vertical direction, by means of a

Fig. 7.

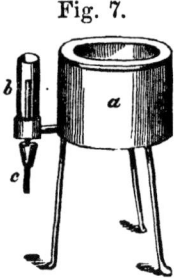

caoutchouc tube; this one is open at both ends, but the fixed tube communicates directly with the interior of the cylindrical vase. Water is dropped gradually into the space between the two tubes, and in this way the vessel, *a*, is filled to the level of the upper opening of the tube *c*, through which the excess is carried off. It is evident that in proportion as the water evaporates, it is replaced by the supply from b, and that one can at will raise or lower the level of the water by moving the tube c.

Several little contrivances have been made for self-regulation of the supply of coal gas to a burner when employed to maintain the heat of either drying-stoves or water-baths, so that, once fixed, the temperature shall be practically constant. These are to be procured at the principal dealers in chemical apparatus.

Apparatus made of Platinum.

In order to preserve articles made of platinum, we must carefully avoid melting caustic potassa, soda, or the alkaline nitrates in them; we must also abstain from heating in them metals or their reducible oxides, or easily reducible salts, such as phosphates, the latter especially in combination with organic matter. We should not heat them strongly and directly over a charcoal fire, still less, one of coke or coal. Platinum crucibles, etc., may be cleaned by friction with fine sand of rounded grains, such as sea sand usually is, or by melting, if necessary, some acid sulphate of potassium, in order to dissolve matter adhering to the surfaces.

Platinum vessels are rapidly injured by the flame of ordinary coal gas—the texture of the metal becomes open, rough, and porous, even though the weight may not alter. For many other agents that must not be brought into contact with platinum vessels, see Fresenius's work on Analysis.

Muffle Furnaces.

The earthenware muffle furnaces furnished by commerce, of French, German, and English make, are generally very suitable; but they have one inconvenience, the opening of the furnace intended to communicate with the opening of the muffle is never sufficiently large to allow the latter to be rightly placed. The result of this arrangement is, that a chink is left, through which the dust of the fuel may penetrate, and falsify the result of an analysis. It would be also an advantage to so arrange that the opening intended for the introduction of the fuel should not be at the same side as that through which the matter to be calcined is introduced. This furnace is heated either with coke or charcoal, in small pieces to insure that the filling take place uniformly and in a regular manner. Every time that a new muffle is employed, care should be taken to wipe it out well lest any of the loose sand detach itself from the surface in the course of an operation, and fall into the capsules or crucibles. In placing these vessels directly on the bottom of the muffle we incur the risk of particles from it or dust from the coal adhering to them; we recommend the avoidance of this evil by placing them on the cover of a porcelain crucible, the ring of which has been previously broken off.

In an analytical laboratory to be kept at work day by day, the muffle furnace had best be regularly built of fire-brick, which has been sanctioned by the great experience of the Cornish and South Wales assayers.

The portable clay furnaces of the French, etc., are very elegant to look at, and very handy for occasional use, but when kept at work from day to day their thin walls get so heated, and radiate so much heat as to be unbearable. Besides this, the same uniformity of temperature cannot be maintained as in the more massive brick furnace; and more muffles are hence lost and wasted by cracking through brusque changes of temperature.

Taking of Samples for Analysis or Assay.

It is obvious that the average composition of a mineral cannot be accurately known by the analysis of a sample taken at random. At the same time it is impossible to operate upon a large quantity at once. To ascertain the average composition as exactly as possible, it is necessary to procure certain portions which, though relatively small in volume, may represent the average composition of the whole mass. From this reduced quantity the specimen for analysis is taken. To obtain this result, we must act according to circumstances; if, for instance, it be a question relative to an entire cargo of mineral just arrived by sea or land, we should set aside at the time of unloading, according to the importance of the mass, or the intrinsic value of the ore, every fifth, tenth, or twentieth load for an especial heap, which may be considered to have the same composition as the whole mass. If the quantity thus set aside be still too large, we may do with this as we did with the whole cargo, and by repeating the process, reduce the quantity to a heap of from one-half to two cubic metres in size.

If the ore be in large blocks we must take care to reduce them to fragments sufficiently small to mix together; selecting small portions here and there we mix them together again, and from this we abstract a small quantity, which we dry at 100° Cent., taking note of the loss it undergoes in the process. We then pulverize in a metallic mortar, and pass the powder through a sieve, the meshes of which are from one to two millimetres in diameter. When the whole has been sifted and a new mixture made, another fractional sample is selected, pulverized finely, and passed through a silk sieve. This is the sample for assay or analysis (*prise d'essai*). In this manner we obtain a mass which represents the average composition of the entire one; but in ordinary cases it answers to select a certain quantity of ore from different places on the heap, observing the proportion existing between the large and small pieces, until we have attained the volume given above, viz., from half to two cubic metres in size, and then proceed as already described. The best test for the accuracy of the result is to repeat the process, analyzing the two final specimens separately. If the operation be well done the analyses should be almost identical.

The "sampling" of copper ores in Cornwall is executed with great care and precision by the assayers, whose custom, however, is rightly never to take samples from heaps of ore uncrushed and in large lumps, as suggested in the text, as being capable of leading to grave errors, or even to fraud. A certain system as to taking the samples and reducing the volume of an average sample is adopted.

The Swansea assayers employ an admirable form of grinding mill for reducing to an impalpable powder the final sample, from which those for assay are taken. It consists of a spheroidal cast-iron mortar or pan which revolves upon a vertical axis. Within it works a cast-iron pestle, to which, by a simple but beautiful train of encyclical wheels, an internal hypocycloidal movement is given, combined with a circular motion round the pan; and at recurrent intervals a rectilinear diametral movement across the pan. The pestle is so arranged that its pressure upon the pan bottom can be regulated at pleasure. The whole is driven by a winch and porter wheel by a man or boy. The cost of one of these mills is said to be about £10, and such would prove a valuable item of apparatus in any laboratory largely engaged on the analysis of ores of iron, especially the harder ores.

PART III.

VOLUMETRIC ANALYSIS OF IRON.

I.—*By means of Permanganate of Potassium.*

THE process for the volumetric determination of iron known as the Marguerite process, from the name of the chemist who discovered it, is based on the following considerations: If to a solution of a proto or ferrous salt of iron, containing a sufficient quantity of free acid, we add a solution of permanganate of potassium, the protosalt of iron (*ferrosum*), is changed to the persalt (*ferricum*), taking from the permanganic acid set free a part of its oxygen, and causing it to pass into the state of a protosalt of manganese, as the following equations show:—

1st phase:

$$K^2Mn^2O^8 + H^2SO^4 = K^2SO^4 + H^2Mn^2O^8;$$

2d phase:

$$H^2Mn^2O^8 + 7H^2SO^4 + 5Fe^2S^2O^8 =$$
$$Mn^2S^2O^8 + 8H^2O + 5Fe^2S^3O^{12}.$$

The salts of iron in the liquid have a very slight coloring power, and that of the salt of manganese is almost nil, that of the permanganic acid, however, is considerable; it follows that as soon as the quantity of permanganic acid added to the solution of iron is sufficient to change the protosalt to the persalt, a very small amount of permanganate in excess will communicate to the liquid a very apparent and persistent rose tint, which will serve as an indication of the end of the reaction. By knowing on one side the volume of the solution of permanganate employed to produce this rose tint, and on the other the quantity of iron that a unit of volume (one cubic centimetre) of this solution can change from a protosalt to a persalt, a simple multiplication will give the quantity of iron contained in the first state in the solution analyzed, and consequently the total amount of iron, if the latter be at its minimum of oxidation.

In application we distinguish three operations: 1st, The preparation of the solution; 2dly, ascertaining its standard*; 3dly, the assay properly so called.

1st. The solution of permanganate must not be either too concentrated, which would impair the accuracy of the method by changing a slight mistake in reading into a serious mistake in the result, nor too dilute, which would be inconvenient from the necessity of filling the burette several times for a single assay. The degree of concentration is correct when the standard of the solution is comprised between 0.010 and 0.007 which corresponds to about 5 to 5.5 grammes of crystallized permanganate to

---

* The *standard* of a solution of permanganate is the quantity of iron that a cubic centimetre of the solution corresponds to, that quantity being expressed in grammes.

the litre of water. The solution is made in the following manner: put all the permanganate to be dissolved into a matrass, add a small quantity of distilled water, and shake the vessel; when the water is nearly saturated, filter the liquid directly into the bottle in which it is to be preserved. The filtration can only be effected through asbestos or gun-cotton, because the permanganate is decomposed by paper or linen. Pour more water on the saline residuum, shake it, and filter again. Repeat the process till all the salt is dissolved; then dilute the liquid with a quantity of water sufficient to obtain the volume which corresponds with the weight of permanganate employed, and shake it strongly to render it perfectly homogeneous. We may substitute with advantage simple decantation for filtration. The solution of permanganate is preserved in a ground stoppered bottle, or in the sort of vessel used for distilled water already described. When the solution is kept in the dark, it will not spoil quickly, provided the permanganate employed be of sufficient purity.

2dly. Ascertaining the standard.—This is effected by iron. Take a piece of this metal, clean and free from oxide, weighing about 0.4 grm.; take the weight very carefully. Put the iron into a matrass of 150 to 200 cubic centimetres in size, with about 50 cubic cent. of distilled water and 4 to 8 cubic cent. of pure sulphuric acid; hydrogen is evolved, and the iron is dissolved, producing the protosulphate of iron—

$$Fe^2 + 2H^2SO^4 = H^4 + Fe^2S^2O^8.$$

As this spoils by contact with the air, it is indispensable to exclude that during the process. Generally speaking, the hydrogen which is liberated during the reaction sufficiently attains this end; but we are more secure by placing in the matrass, along with the iron, a small quantity of carbonate of sodium or potassium and closing the mouth with a caoutchouc stopper, pierced with a slender tube, or a stopper with a valve like that represented by Fig. 8.*

Fig. 8.

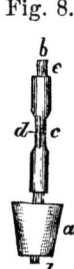

After complete solution (which can be hastened by using moderate heat), we pour the liquid into a cylindrical vessel, rinse out the matrass with distilled water, adding the rinsings to the liquid, and diluting still further with cold water until we get approximately a total volume of half a litre.

We then pour the solution which is to be standardized into a graduated burette, with a caoutchouc

* The play of this stopper is very simple: *a* is a caoutchouc stopper, perforated with the glass tube *b b;* *c*, a caoutchouc tube having a small longitudinal slit, *d;* the tube is closed at its extremity, *e*, with a stopper of glass rod.

clip-cock ;* bringing the top surface of the solution exactly to the zero of the graduation, and placing the vessel containing the solution of iron below it, and beneath that a white surface, a porcelain slab, a sheet of white paper, or a common plate, we pour into it from the burette cock the permanganate, first in a little continuous stream, then intermittently, and finally drop by drop, taking care to stir it constantly with a glass rod until it turns rose-color, and keeps the color after agitation. This coloring is due to a slight excess of permanganate, the amount of which we can ascertain by dropping a little of the "chameleon" solution into another vessel quite similar to that used in the experiment, but containing, instead of a solution of iron, an equal volume of distilled water; we note the fraction of a cubic centimetre necessary to obtain a coloration identical with that of the solution of iron.

Let n be this fraction, and N the number of cubic centimetres employed in the first operations. $N - n$ will evidently be the number of cubic centimetres employed in peroxidizing the quantity of iron contained in the liquid; this quantity is equal to $\frac{996}{1000}$ of the weight of the iron wire dissolved, since the latter contains on an average $\frac{4}{1000}$ of impurities.

---

* The burettes of Gay-Lussac or of Geissler have been recommended, or those provided with a glass cock. The action of the caoutchouc clip-cock upon the permanganate has been feared, but our experience proves that these fears are much exaggerated, and we recommend the simple burette with a caoutchouc spring clip-cock as the most convenient and economical. The glass cocks get frequently stuck fast, or are broken by quick movements.

The standard of the liquid, that is, the weight of iron indicated by each cubic centimetre, is then

$$T = \frac{P\ 0.996}{N - n}$$

A numerical example will complete what has been stated: Let 0.3916 grm. be the weight of the iron wire employed; let us suppose that into the solution of these 3916 ten-thousandths of a gramme of iron we must drop, in order to obtain persistent coloration, 42.7 cubic centimetres of the solution of permanganate, the standard of which we are to determine; if we have ascertained by a comparative experiment that the coloring is due to an excess of 0.2 c. c., the number of cubic centimetres employed in the peroxidation will be $42.7 - 0.2 = 42.5$; consequently the standard sought is

$$T = \frac{0.3916 \times 0.996}{42.7 - 0.2} = \frac{0.3900}{42.5} = 0.0091764\ +$$

The standard thus obtained is to be noted down with the date. We should not be satisfied with one experiment, but make at least two, or even more, if the first two do not give sufficiently concordant results.

It has been suggested that the double proto-sulphate of iron and ammonium (Fr. Mohr), oxalic acid (Hempel), or ferrocyanide of potassium (Gintl) may be substituted for the iron. Ascertaining the standard by means of these substances as furnished by commerce may occasion errors, which for the double sulphate might amount to 2 and 3 per cent.

The necessity of preparing, or, at all events, of purifying these substances renders their employment of little practical advantage in an industrial laboratory, where, in our opinion, iron is the only suitable substance for fixing the standard. Nevertheless, we shall give the methods of operating with the salt of Mohr and with oxalic acid. Weigh 1.5 to 2 grammes of the first, or 0.4 gr. to 0.5 gr. of the second; dissolve the weighed salt in distilled water, dilute the solution to the volume of half a litre, add some sulphuric acid, then drop in some solution of permanganate, until persistent coloring is obtained. With oxalic acid the coloring is very slow at first; it may be accelerated, without altering the results, by adding to the acid solution a small quantity of a solution of the protosulphate of manganese.

The formula of the double sulphate of iron and ammonium is $FeN^2H^8S^2O^8 + 6H^2O$; its molecular weight is equal to 392, which is exactly 7 times that of iron, of which it consequently contains $\frac{1}{7}$. Thus

$$T = \frac{P'}{7\,N - n}$$

P' indicating the weight of the sulphate employed, N and n having the same signification as before. The action of oxalic acid on the permanganate is represented by the equation—

$$5H^2C^2O^4 + K^2Mn^2O^8 + 3H^2SO^4 = K^2SO^4 + 2MnSO^4 + 5H^2O + 10CO^2.$$

Comparing this equation with that representing the action of protosalts of iron on permanganate, we re-

8

mark that one molecule of oxalic acid decomposes as much permanganate as two of the protosalt of iron. The molecular weight of crystallized oxalic acid ($H^2C^2O^4 + 2H^2O$) is 126; double the molecular weight of iron is 112; these two numbers are in the ratio of 9 to 8. If we use a weight p'' of crystallized oxalic acid to ascertain the standard, the latter will be given by the equation

$$T = \frac{8p''}{9\ N - n}$$

Permanganate of potassium is a very delicate reagent, and in experimenting we must carefully avoid any disturbing influence, if we wish to obtain accurate results. We must, therefore, avoid the presence of hydrochloric acid and chlorides. Hydrochloric acid decomposes permanganic acid, producing the protochloride of manganese, water, and chlorine—

$$H^2Mn^2O^8 + 14HCl = Mn^2Cl^4 + 8H^2O + 10Cl;$$

chlorides have the same effect in presence of sulphuric acid.

This reaction goes on along with the reaction of the permanganate on the protosalts of iron, and so falsifies the result; for the chlorine set at liberty does not act upon the protosalts to change them into persalts, or, at least, only acts partially, so that one easily perceives the odor of the chlorine which is evolved when experimenting under such circumstances.

We should also avoid the presence of nitric acid, or any of the nitrates, which, during the reduction of the persalt of iron by zinc, might create nitrogenous combinations less oxygenated than nitric acid, and possessing, like the protosalts, the power of decoloring the permanganate. Certain organic matters also possess this property; and if the substance to be analyzed contain such, they must be got rid of by a previous operation, consisting in the roasting or fusion with sodic carbonate of potassium. The ores (carbonates) of iron from coal formations (*black band*, etc.), contain organic matters.

The zinc used for the reduction of the persalts of iron should be completely dissolved or abstracted from the liquid before proceeding with the assay, because though it does not act upon the permanganate either alone or by the hydrogen which it liberates, as might be supposed, yet it falsifies the results, by partially reducing the salts of iron at the same time as the permanganate oxidizes them. Mitscherlich is of opinion that the zinc should be entirely and previously dissolved, because occasionally that metal precipitates iron in an appreciable quantity.

The solution of the substance to be assayed must contain a sufficient proportion of free acid (sulphuric acid), without which the permanganate is only partially reduced, and the liquid, being rendered turbid by the precipitation of a certain quantity of the peroxide of manganese, becomes brown; in which case the experiment has completely failed, and must be recommenced. Lastly, we must take care to operate always with liquids which are not sensibly warm, and which in all cases occupy approximately the same volume; and we must not forget to subtract from the number of cubic centimetres employed the necessary volume for the excess in coloring, and also that corresponding to the quantity of zinc used for reduction, as we shall see further on.

3dly. The assay is made on one gramme of material reduced to fine powder, in case it be an ore; the solution is made by means of 4 to 6 cubic centimetres of concentrated hydrochloric acid, digested for half an hour with gentle heat; after which, if the ore be sufficiently attackable by hydrochloric acid, the insoluble residue (silica and clay) will appear very white. If that be the case, add 5 to 6 cubic centimetres of sulphuric acid, and heat again to get rid of the hydrochloric acid, the presence of which, as we have already said, should be avoided; the chlorides are thus changed to sulphates. When the ore is only partially attackable by hydrochloric acid, we fuse it with three or four times its weight of sodic carbonate of potassium in a platinum crucible, and then treat the fused mass with warm dilute sulphuric acid until completely disintegrated. We might also treat it with dilute hydrochloric acid, but in this case the solution must be evaporated to dryness after the addition of sulphuric acid.

In whatever way the solution has been prepared, it must be diluted with cold water to obtain a volume of 150 to 200 cubic centimetres; we then immerse in it a slip or two of zinc, to reduce the persulphate of iron to the protosulphate—

$$Fe^2S^3O^{12} + Zn = Fe^2S^2O^8 + ZnSO^4.$$

This reaction is not the only one produced; the excess of sulphuric acid acts in its turn upon the zinc, and liberates hydrogen—

$$H^2SO^4 + Zn = ZnSO^4 + H^2.$$

When the liquid is too diluted, the reduction is slow; but on the other hand, if it be too concentrated, the solution of zinc is too irregular or too rapid: the proportion we have given may be considered a just mean between these two extremes. During the reduction, contact with the air must be avoided; this result is obtained by providing the matrass in which the operation goes on with one of those valve-stoppers before described when treating of ascertaining the standard, or by covering it with a watch-glass. The liberated hydrogen is then sufficient to exclude the air from the matrass; and it is thus unnecessary to use the carbonate of potassium or sodium, as in the other case. The liquid, which is at first yellow, gradually loses its color, and when this is complete, the reduction is so. When we wish to be certain of this, pour a drop of the liquid into a small quantity of the solution of sulphocyanate of potassium; if no red color be immediately produced, the reduction has been complete. The solution of the zinc is then effected by the addition of sulphuric acid, and we collect together the flakes of black matter floating in the liquid by gently shaking; we allow these to deposit, and then decant into a cylindrical jar, washing the residuum two or three times, and adding the rinsings to the first liquid decanted. These operations being performed as rapidly as

8*

possible, we dilute with cold distilled water, until
we get the volume used in all our volumetric assays,
viz., about half a litre.   In order to prepare the
solution of the ore for the experiment, it is sufficient
to add sulphuric acid, if it does not already contain
enough.   We then drop the solution of permanganate from a graduated burette, with the precautions
indicated with respect to the standard, until the
persistent rose-color appears.   By the graduated
scale of the burette, we learn the volume of the
solution of permanganate employed, and, knowing
its standard, we find by simple multiplication the
quantity of iron contained in the weight of the substance submitted to examination, and consequently
the value of the ore.   The reduction previous to this
analysis is effected by the help of zinc, as already
stated; mercantile zinc is generally employed, and
this, unfortunately, always contains a certain amount
of iron; although the proportion be small, we are
obliged to allow for it, on account of the considerable
weight of zinc employed and the erroneous results
it would therefore lead to.

We can easily correct this error by using zinc cut
into pieces of equal size from the same sheet, and
determining the fraction of the cubic centimetre of
permanganate necessary to peroxidate the iron contained in one of these pieces.   This fraction, or one
of its multiples, if several pieces of zinc have been
required for the reduction, must be subtracted from
the number of cubic centimetres employed for the
assay before converting, by calculation, this volume

into the weight of iron sought. We must not forget also to subtract the excess of permanganate required to produce the coloring—an excess ascertained by the process already described.

II.—*By means of Protochloride of Tin.*

The protochloride (*stannosum*) of tin possesses the property of reducing the persalts of iron to the state of protosalts at a temperature of nearly 100°, as the following equation shows:—

$$2Fe^2Cl^6 + Sn^2Cl^4 = 2Fe^2Cl^4 + 2SnCl^4.$$

From this reaction it follows that by pouring a solution of protochloride of tin of known standard into an iron solution until the latter be completely reduced, we can ascertain from the volume of the solution of tin employed, the amount of iron in the condition of a persalt contained in the liquid assayed. In order to fix the completion of the reaction, it has been proposed to add to the liquid a few drops of the sulphocyanate of potassium, which produces a deep red color with persalts of iron; but the sulphocyanate being itself a reducing agent, under the conditions in which it is necessary here to operate, its presence as an indicator might interfere with exact results.

To avoid this it is necessary to slightly complicate the process by proceeding as follows: Pour into the solution of iron such a quantity of protochloride of tin as is sufficient to completely dissipate the yellow color of the persalts, which, in an acid solution when heated, is very intense. To obtain this result we generally must add a slight excess of the reagent; this excess is determined by means of a solution of iodine, which saturates the tin, and the least excess of which is ascertainable by starch, which becomes of a deep blue.

The probable reaction between the protochloride and the iodine is either

$$Sn^2Cl^4 + 4I = SnCl^4 + SnI^4,$$

or

$$Sn^2Cl^4 + 4I = 2SnCl^2I^2.$$

As the solution of the protochloride of tin alters very rapidly, it is necessary to ascertain its standard before each series of experiments is undertaken; this is easily done by ascertaining the number of cubic centimetres necessary for the reduction of a given quantity of a standard solution of perchloride given.

We have, therefore, three solutions to prepare:—

1st. A solution of the protochloride of tin ;

2d. A solution of iodine;

3d. A solution of the perchloride of iron.

The last only requires to be standardized; it is of suitable concentration when it contains 0.01 gramme of iron for each cubic centimetre of volume. This particular standard has, moreover, the advantage of simplifying further calculations.

In order to obtain this solution in the easiest manner, dissolve an ascertained weight of pure iron in hydrochloric acid, add to this solution the quantity of chlorate of potassium required to change the protochloride into the perchloride of iron, and boil until free chlorine is completely expelled, then add of water sufficient to make the number of cubic centimetres correspond with the number of centigrammes contained in the weight of pure iron. In our calculation we must not forget that, as before stated, the purest commercial iron contains on an average only 0.996 of iron.

In default of metallic iron we may employ any solution of the perchloride of iron, determine its exact standard, and dilute with the requisite quantity of distilled water.

The solution of tin is prepared by dissolving 10 to 20 grammes of crystallized protochloride of tin in a litre of distilled water, strongly acidulated with hydrochloric acid. We may also obtain a suitable solution by dissolving 5 to 10 grammes of grain tin in hydrochloric acid with heat, and diluting the solution to about a litre in volume. Filter, if necessary, in order to obtain a perfectly limpid solution, and preserve in a ground stoppered bottle.

The solution of iodine is made by dissolving about 5 grammes in a litre of distilled water; if the solubility of the iodine in water be too small for this proportion, we can increase its solubility by adding a small amount of iodide of potassium.

We should also prepare a dilute solution of starch by boiling a small quantity of starch with water.

The progress of the operation is as follows:—

1st. Determine the relation between the solution of tin and the solution of iodine; for this purpose

take one or two cubic centimetres of the first, drop
in from a glass three or four drops of starch solution,
then add, if it be not already very acid, one or two
cubic centimetres of hydrochloric acid diluted with
its volume of water; then pour in from a graduated
burette the solution of iodine until the appearance
of the blue color; note the number of cubic centimetres employed to reach this point, which shows
that the reaction has been completed.

2d. Determine the quantity of solution of tin
necessary to reduce a known weight of iron. This
is done in the following manner: Take ten cubic
centimetres of the solution of iron containing 0.1
gramme of iron, add 10 cubic centimetres of dilute
hydrochloric acid, heat the whole in a matrass to
ebullition. Then pour in slowly from a graduated
burette some solution of tin until the disappearance
of the yellow color; cool by agitating the matrass
in cold water, add some drops of starch solution, and
determine the quantity of solution of iodine necessary
to saturate the excess of chloride of tin which has
been used. The latter is calculated by means of the
proportion furnished by the first operation, having
regard to the number of cubic centimetres employed.

3d. We now proceed to the assay properly so
called. For this purpose dissolve a gramme of the
material to be assayed in a small matrass with
hydrochloric acid, convert all the iron into the state
of perchloride by the addition of a small quantity of
chlorate of potassium, and boil to get rid of the free
chlorine. Pour some solution of tin into the boiling
liquid until the reduction be complete, without

which it would be necessary to separate the insoluble matter by filtration. Finally we determine the excess of the chloride of tin by the method already described with the solution of iodine. By comparing the number of cubic centimetres of the chloride of tin employed for the operation with that required for the reduction of 0.1 grm. of iron, we calculate the quantity of this metal contained in the substance assayed. Thus, for example—

To a cubic centimetre of the solution of the proto-chloride of tin, acidulated with hydrochloric acid, we have added starch and poured in solution of iodine; 1.4 cubic centimetre of the latter was required to obtain the blue color. We then took 10 cubic centimetres of the standard solution of iron, poured into it 19.2 c. c. of the solution of tin, and then 0.7 c. c. of the solution of iodine. These 0.7 c. c. of iodine correspond, according to the first experiment, to 0.5 c. c. of chloride of tin; the volume of the latter required for the reduction of 0.1 gramme of iron was therefore 18.7 c. c. Again, having dissolved one gramme of the iron ore with the prescribed care, and having allowed for the excess of the chloride indicated by the iodine, we found that 42.3 c. c. of the solution of tin was required to obtain the complete reduction.

The proportion $18.7 : 0.1 :: 42.3 : x$, gives the quantity of iron contained in one gramme of the ore; this is—

$$\frac{42.3 + 0.1}{18.7} = 2.2673 +$$

The ore contains, therefore, 2.2673 per cent. of iron.

A third volumetric method for the determination of iron not noticed in the text, yet should not be wholly passed over—viz., that introduced by the late Dr. Penny, of Glasgow, and much employed by him for the assay of iron ores. This consists in the employment of a standard solution of bichromate of potassium, in place of the permanganate of Marguerite.

This process depends upon the fact that an acid solution of a protosalt of iron, acted upon by one containing chromic acid, is changed to a persalt of iron, and the chrome to a sesquisalt, so that the reaction of bichromate of potassium and protochloride of iron is thus :—

$$6FeCl + 7HCl + KO2CrO^3 = 3Fe^2Cl^3 + KCl + Cr^2Cl^3 + 7HO,$$

as given by Percy (Metallurgy of Iron), who recommends this method as preferable to Marguerite's. The standard solution advised by Percy is of 305 grains of bichromate in 35,000 grains of water, 1000 grains of such solution being equivalent to 10 grains of iron. The iron solution is brought to the state of a protosalt by boiling either with zinc (of known degree of purity) or with sulphite of sodium ; and the completion of the assay is determined by a test solution of red (ferricyanide of potassium) formed by two or three grains of the salt to half a pint of water. Dr. Percy gives minute instructions (as above) for the application of this method, with the successful use of which organic matter in the iron ores does not interfere chemically, though it may optically, by remaining in suspension and obscuring the color indications.

On the strength of some years' experience of both methods, Percy prefers this to that of Marguerite on the ground that bichromate of potassium is readily obtained in crystals in commerce, and needs not to be specially prepared ; that the solution is easily prepared ; that once standardized the solution is much less liable to change, and therefore more reliable than that of permanganate of potassium ; that it is less liable to error of manipulation from evolution of chlorine, or from organic matters present; and that, while equally accurate, it occupies less time than the permanganate process. The less alterable character of the standard bichromate solution is certainly an advantage. On the other hand, however, it should be remarked that at present (if not in 1864, when Dr. Percy wrote) permanganate of potassium can be obtained commercially in crystals of great purity, whilst the purity of commercial bichromate of potassium has seriously declined latterly, owing to the process of calcination of the chrome iron with lime instead of with nitre or alkali, as formerly practised by the manufacturer. The result is that all the commercial bichromate which the writer has come across of late years, contains a considerable and variable amount of sulphate of lime (perhaps in the state of some double salt), but crystallizing out in colorless small tabular crystals of the normal form of sulphate of lime. Hence to obtain at present reasonably pure bichromate for the standard solution, we must submit the salt of commerce to at least two successive crystallizations.

In general, on the subject of volumetric assays of iron, the writer ventures to remark that the assayer who has not mastered the general principles of this method by the study of Sutton's, or some other of the standard works on Volumetric Analysis, will find himself, when dealing with compounds of iron new to him, very liable to slip into errors that may be to him inexplicable ; that the successful use of the volumetric method demands a large amount of very special, though narrow, experience and practice ; that as dependent on slight differences of color, the results of different experimenters are subject to all the differences in the appreciation of colors appertaining to different eyes. It becomes more and more probable not only that more people are color blind than is commonly supposed, but that to no two people does the same color appear precisely alike.

Lastly, the value of volumetric assay for saving of time is a function of the largeness of the number of, *chemically*, *quite similar* substances requiring to be assayed. Where these are not numerous, or where there may be chemical differences, such as to disturb the color indications, or to introduce any chemical uncertainty requiring afterwards to be sought out, other and older methods dependent not on measuring volumes, but on the balance, are to be preferred.

[We recommend again the entire solution of the zinc employed for the reduction of the persalts of iron, because, as already pointed out by Mitscherlich, we have also found that zinc occasionally precipitates a sensible quantity of iron.

As titaniferous iron ores are quite abundant in certain parts

9

of the United States and Canada, and are coming into use, we here give the volumetric methods of analyzing them.

Zinc cannot be employed for reducing solutions of titaniferous irons, because the perchloride of titanium is also reduced to the state of sesquichloride, and will afterwards be peroxidized by the solutions of permanganate of potassium or of bichromate of potassium. The reduction of such solutions should be effected by means of the sulphurous anhydride or the sulphite of sodium, and the liquor should be boiled until all sulphurous smell has disappeared. By these operations, unless the liquor be very acid and concentrated, a part of the titanic acid will be precipitated, in the shape of a white powder. In this case it is preferable to employ the Penny's process by bichromate of potassium, because the slight pink coloration by the permanganate will be ascertained with difficulty.

Another method of volumetric analysis, convenient for titaniferous and other iron ores, consists in employing a solution of persalt of iron, which is transformed to the state of protosalt by a standard solution of hyposulphite of sodium.

For instance, let us standardize a solution of 25 grammes of hyposulphite of sodium in 1 litre of water. One gramme of piano-wire, supposed to contain $\frac{996}{1000}$ of pure iron, is dissolved in hydrochloric acid, and then peroxidized by chlorate of potassium, chlorine, or bromine, but not with nitric acid. The excess of chlorine or bromine is removed by boiling, and the solution is diluted with water so as to occupy 100 cubic centimetres.

We put 20 c. c. for instance, of this solution in a beaker, and add some hydrochloric acid and 1 or 2 drops of a solution of sulphate of copper (which aids in the transfer of oxygen from the iron to the hyposulphite). The whole is colored a dark-red color by a few drops of a solution of sulphocyanate of potassium.

The solution of hyposulphite of sodium is then poured from a graduated burette until the red coloration disappears, that is, when the whole persalt of iron is transformed into a protosalt. From the number of c. c. of hyposulphite of sodium employed for the iron contained in the 20 c. c. we obtain the standard of the liquor.

The operation is performed in the same manner for the analysis of the iron ore. The solutions should be rather concentrated, and contain a certain amount of free hydrochloric acid, which is best arrived at by practice. A temperature of about 40° C. aids the reaction. The proportion of 25 grammes of hyposulphite of sodium for 1 litre of water, gives a solution of convenient concentration, which should be standardized almost every other day, because it is easily altered.

Our experience of this process is, that it is satisfactory for practical purposes, when the standard solution of hyposulphite of sodium is freshly made. With an old one, the reaction lags behind, and the change of color is not rapid or very distinct.]

PART IV.

——◆——

ANALYSIS OF IRON ORES.

Iron ores may contain, in addition to the oxides of iron, water, carbonic anhydride, organic matters, silica, alumina, lime, magnesia, phosphoric acid, sulphur, arsenic, zinc, lead, and manganese, in various degrees of oxidation. In rare cases they may also contain other matters; we shall limit our description to the processes requisite for the quantitative analysis of the elements which enter into the composition of ordinary ores, referring for special cases to larger analytical treatises.

Before commencing the analysis of an ore, it is very important to know whether it contains manganese or not; in fact, the process to be pursued in each case is different, dependent upon this.

We can quickly ascertain the presence of this metal by fusing the ore with two or three times its weight of carbonate of potassium; the fused mass assumes a green color, more or less dark, according to the amount of manganese. The blowpipe will usually suffice for this.

[The presence of manganese in an ore may also be ascertained by the wet method, as follows:—

Heat with diluted nitric acid a certain quantity of dioxide of

9*

lead, and pour into it a few drops of a nitric solution made with
the ore on trial.   If manganese be present, the supernatant
liquor will be colored a purple-red, the more intense as the proportion of manganese is greater.

Or the dioxide of lead and the powdered ore may be heated
together in the same vessel with nitric acid, diluted with about
twice its volume of water.   The mixture is stirred now and then,
and the coloration will appear when the solid particles have
subsided.

The blowpipe assay is best made upon a platinum foil or spoon,
and is aided by the addition of a small proportion of nitrate of
potassium to the carbonate of sodium (or potassium) used.]

I.—ORES NOT CONTAINING MANGANESE.

The analysis of these ores is conducted as if they
only contained silica, alumina, peroxide of iron, lime,
and magnesia; the determination of other substances
is made with special experimental specimens, either
to facilitate the operations generally, or because the
proportion of such substances is so small that to
ascertain the amount with any degree of accuracy it
is necessary to operate with quantities relatively
large.

Weigh a gramme of ore, and if it contain organic
matter, the presence of which might be inconvenient
in the course of the operation, destroy it by roasting
or still better by fusion with three to four parts of
sodic carbonate of potassium, and half a part of
saltpetre.   When we operate by roasting, a second
weighing of the specimen gives directly the loss
sustained by the process.   This loss is generally
ascertained with a special portion because calcining
renders mineral ores more refractory to the action of

acids. It is to avoid this inconvenience that we prefer generally to destroy organic matter by fusion.

When it is unnecessary to have recourse either to roasting or fusion, we pour some cubic centimetres of hydrochloric acid upon the ore, and heat gently in a closed flask; when the acid has acted for a little time we add a little nitric acid, the object of which is to facilitate the solution and to transform any protosalts of iron present into persalts. After digesting for half an hour or an hour, the insoluble remainder ought to appear either as a gelatinous mass or a white powder, at most a little grayish. When neither of these results follow, the ore is refractory to the action of acids, and it is necessary to fuse it with the alkaline carbonates. In this case we melt a gramme of the substance with four to five grammes of sodic carbonate of potassium in a platinum crucible, covered to avoid the spirting from the liberation of carbonic anhydride, expelled by the silicic acid. We raise the temperature gradually to a red heat, and keep it at or above this point till the mass is in tranquil fusion; then cool it rapidly. After cooling, the crucible and cover are placed in a deep dish of Berlin porcelain, containing enough water to cover them both. We then heat, taking care to keep the water below ebullition, and to add in succession very small quantities of hydrochloric acid, until the fused mass is completely disintegrated. We must follow exactly the same plan if the object of the fusion had been merely the destruction of contained organic matter. The solution of an ore incompletely soluble in acids may be effected by submitting it first to the action of hydrochloric acid only; when this action has ceased dilute the liquid with at least its own volume of water, that it may not attack the filtering paper; filter and wash the residuum, dry it in the filter, and roast both together. Add to the product of the roasting, and in the platinum crucible which has served for the operation, three to four grammes of sodic carbonate of potassium, stir the whole well with a platinum wire or glass rod, and fuse the mixture, and after fusion heat it in the manner and with the same precautions we have indicated with the acid and hot liquid obtained by filtration after the first reaction, and which in this case is substituted for pure acidulated water.

The acid solution of the ore obtained by either method is carefully evaporated to dryness in the water bath in a porcelain capsule, by which the silica is rendered insoluble in acids. The residue of the evaporation is well moistened with concentrated hydrochloric acid, and then treated with hot water, until it is completely disintegrated, a process which may be hastened by gently crushing it with a perfectly smooth agate pestle. We wash the insoluble residuum by two or three decantations, and finally throwing it on the same filter employed for filtering the washing waters, we continue to wash it with boiling water, to which we add at the commencement a little hydrochloric acid. The insoluble matter is first dried, then roasted, and finally let cool under the drying bell. If in drying the edges of the filter turn yellow, it is an indication that the washing has not been sufficient, and in this case we should recommence the analysis. If we have made use of the alkaline carbonates in order to make the ore soluble, the silica after roasting should be perfectly white, and should crumble into powder; but if the ore has been simply treated with acids, aluminous matter may be mixed with the silica, and in this case the latter is not so white, and is more or less coherent. Fusion with the alkalies is, therefore, indispensable if we require to know exactly the amount of silica and alumina which enter into the composition of the ore.

In the liquid separated from the silica by filtration we can precipitate alumina and peroxide of iron by a slight excess of ammonia,* the ammoniacal liquid is heated till the precipitate be well reduced in bulk; this precipitate contains, in addition to alumina and peroxide of iron, phosphoric acid, when operating with phosphoric ores. We must filter rapidly to prevent the ammonia absorbing carbonic acid from the air, and wash the precipitate in warm water, first by decantation, and then on the filter, until a drop of the washings filtered through, and gently evaporated on a sheet of platinum by a Bunsen burner or spirit-lamp, leaves no residuum. When the precipitate is well washed it is dried, roasted, and weighed, and from its weight we deduce that of the alumina; this is done by subtracting from the total weight

* This reagent should be free from carbonate to avoid precipitating lime. This condition is important, for magnesia and lime, which are slightly precipitated along with oxide of iron even when the ammonia is pure, would be much more so if the ammonia was partially carbonated.

that of the peroxide of iron, which is to be obtained
by a volumetric assay, and the weight of the phosphoric acid furnished by an especial assay for the
same. The liquid from which the peroxide of iron
and the alumina have been withdrawn now contains
lime and magnesia; to separate these two bases, and
ascertain their amounts, we must precipitate the
lime by means of oxalate of ammonium, of which
we add an excess, in order to maintain the solution
in the state of a double oxalate, the oxalate of
magnesia being but slightly soluble.

It is advisable to effect the precipitation of the
oxalate of calcium from a warm solution, and to
allow twelve hours for the formation and deposition
of the precipitate. After this lapse of time we pour
the clear solution on a filter, and cause the precipitate to collect on it by means of a jet of warm water;
if some of the precipitated oxalate adhere to the
sides of the precipitating vessel, it is necessary to
redissolve it with a drop of nitric acid, and precipitate it again with ammonia. We add this second
precipitate to the first, and subject the whole to the
washing process. To obtain the weight, we calcine
the oxalate by heating it progressively to clear redness, maintaining this temperature for a quarter of
an hour; by this treatment it is first changed into
the carbonate, and at last into anhydrous or caustic
lime; it is cooled under the drying bell, and when
cold it is weighed as rapidly as possible to prevent
its absorption of moisture and carbonic acid from the
air. After weighing, we can ascertain its complete
transformation into lime by pouring on the precipi-

tate as much water as will cover it, and adding a drop of hydrochloric acid. If the decomposition is complete, there should not be the least liberation of gas. If we find that the calcination has been insufficient, instead of repeating the experiment, we can weigh the lime as a sulphate; for this purpose pour into the crucible some drops of sulphuric acid, evaporate with care to dryness, calcine at a low red heat, and weigh after cooling in the drying bell.

The standard solution made use of in the volumetric assay of iron, viz., that of permanganate of potassium, may very well be employed also for the volumetric assay of lime. This assay is made in the following manner: Redissolve on the filter the well-washed precipitate of oxalate of calcium in the least possible quantity of slightly diluted hydrochloric acid, wash the filter, and, adding to the liquid a little solution of protosulphate of manganese and sulphuric acid, pour in some solution of permanganate (with the same precautions as indicated for the determination of iron), until the appearance of a slight, persistent rose-color. The quantity of lime indicated by one volume of the standard solution is exactly equal to half the quantity of iron corresponding to the same volume. The reaction of the permanganate on the oxalate of calcium in presence of an excess of acid, is represented by the following equation:—

$$5CaC^2O^4 + 2KMnO^4 + 8H^2SO^4 =$$
$$10CO^2 + K^2SO^4 + 2MnSO^4 + 5CaSO^4 + 8H^2O.$$

The addition of the protosulphate of manganese to the liquid is intended to hasten the action of the permanganate, and admit of the operation being performed cold, which, in addition to the fact that the liquid contains hydrochloric acid, make this indispensable. To obtain the quantity of magnesia, add to the liquid separated by filtration from the oxalate of calcium the third of its volume of ammonia, then an excess of phosphate of sodium; stir well and leave at rest in a covered vessel from twelve to twenty-four hours. The precipitate should be made in a perfectly unscratched glass jar, and in stirring the liquid care should be taken not to touch the sides of the vessel with the glass rod; otherwise portions of the crystalline precipitate of the double phosphate of ammonium and magnesium which is formed would adhere to the sides of the jar, so as to be difficult to detach; should this be the case we must redissolve with hydrochloric acid and precipitate with ammonia.

The precipitate is thrown on a filter, washed with cold water containing the third of its volume of ammonia, then dried, calcined, and weighed. In calcining it loses ammonia and water, and is changed into the pyrophosphate of magnesium—

$$2Mg(NH^4)PO^4 = H^2O + (Mg^2P^2O^7) + 2NH^3.$$

If we calcine the double phosphate too rapidly, the resulting mass is compact, often gray and even black in the interior. According to Rose, this results from the presence in the precipitate of small filaments of the filter or other organic matters, the carbonization of which occurred before the liberation of the ammonia. To avoid this, we should raise the temperature slowly to redness; in this way the precipitate becomes pulverulent, and the organic matters are completely burned away.

There are various precautions necessary to success in the separation of lime and magnesia, for which the assayer should consult the larger works on Analysis (Rose, Fresenius, etc.). In particular, Scheerer has pointed out that when the lime present bears a small proportion to the magnesia, the former goes down with the latter wholly or in part; and he has proposed in such cases to convert both bases into sulphates, and separate the sulphate of calcium by alcohol. Sonstadt has proposed and described in detail a method of separating these bases by tungstate of sodium, which appears valuable. As regards the separation of calcium from aluminium in solution, the remarks of H. Rose also should be known. By his method ammonia absolutely free from carbonate is not indispensable.

Instead of determining the oxide of iron volumetrically, and the alumina by difference as we have described, we may determine the amount of each by weighing, alumina being soluble in a solution of caustic potassa, a property which oxide of iron does not possess. The precipitate obtained by ammonia is redissolved on the filter with the least possible volume of hydrochloric acid; this solution is then dropped into a moderately concentrated solution of caustic potassa, and heated to boiling in a porcelain dish. We must take care to stir constantly while mixing the two liquids. To avoid using too large a quantity of caustic potassa, which, when pure, is an expensive reagent, we may previously saturate the most of the excess of acid used for the solution of the iron and alumina with carbonate of potassium or of sodium. After treating the liquid with the caustic potassa, we proceed to filtration, taking care to dilute the

10

alkaline solution with enough water, that it shall
not corrode the filter; in this manner the insoluble
oxide of iron is separated. The alumina is again
precipitated with ammonia, or still better with the
hydrosulphide or the carbonate of ammonium, after
the liquid has been slightly acidulated with hydrochloric acid. When the alumina has been thus precipitated, it is thrown upon a filter, washed carefully,
dried and weighed after calcination.

The peroxide of iron left on the filter cannot be
directly weighed; it always retains, notwithstanding
the washing, a certain amount of alkali; to get rid
of this it is dissolved in hydrochloric acid, and precipitated again, the solution being *hot*, with ammonia.
The oxide thus obtained is filtered out, washed, dried,
calcined, and weighed.

[The preceding directions may be sufficiently accurate for
every-day work, in iron works, for instance; but, in analyses
requiring a great deal of care, as for the determination of the
formula of a mineral, it will be found that the separations of the
various constituent parts are not so thorough as may be desired,
and that further work is required.

For instance, in ores holding lime and magnesia, it is nearly
impossible to have the precipitate of oxide of iron and alumina
free from these substances, by one operation, notwithstanding
care has been taken to add the ammonia by small quantities at a
time, in boiling liquors, and only in slight excess, that is to say,
with the same precautions recommended in the following pages
for the separation of manganese by ammonia.

The precipitation of the lime and magnesia with the oxide of
iron will more readily take place in concentrated liquors than in
diluted ones. If the proportion of iron be small, comparatively
with that of lime, the precipitate by ammonia may be more white
than brown.

The presence of a certain amount of chloride of ammonium in

the liquors, which is added, or formed by the neutralization of the hydrochloric acid by ammonia, aids in retaining the manganese and magnesia in solution.

Therefore, when accuracy is desired, the still moist precipitate of iron and alumina is, after washing, dissolved upon the filter with hydrochloric acid, and reprecipitated by ammonia, with all the necessary precautions. It is sometimes necessary to repeat this operation two or three times, if we need great accuracy.

The precipitation of the alumina and iron by the hydrosulphide of ammonium, allows of a more rapid and complete separation from the lime and magnesia, and the best mode of operation consists in passing a brisk current of hydrosulphuric acid through the solution, and adding small quantities of ammonia at a time. Very little, if any, carbonate of lime or of magnesia will become precipitated. The liquors had better be in a flask which may be corked, so that contact with the air is prevented when the alumina and sulphide of iron are allowed to settle. When the supernatant liquor has become yellow, without any green in it, it is decanted and filtered, and the precipitate is washed again once by decantation, and afterwards upon the filter, in a covered funnel, taking care to operate as rapidly as possible and to have some hydrosulphide of ammonium in the washing liquor, on account of the rapid oxidization of the sulphide of iron. The washed precipitate is immediately dissolved upon the filter with hydrochloric acid, and the solution is heated to expel the hydrosulphuric acid which may be present. The iron is then peroxidized by chlorate of potassium, or nitric acid, or bromine, and the oxides are precipitated by ammonia. After a thorough washing by decantation and upon the filter, the precipitate may be considered free from lime and magnesia.

It is proper, on account of the lime and magnesia, to precipitate with but a slight excess of ammonia, and, if too much of this reagent has been added, to continue the boiling until the ammonia smell has disappeared. On the other hand, a large excess of ammonia has the advantage of decomposing to a certain extent the insoluble subsalts of iron formed during the precipitation. Therefore, the last precipitation will be made with but a slight excess of ammonia, and, after decanting and filtering the supernatant liquor, the precipitate will be washed once by decantation with boiling water. It will be afterwards boiled with ammonia and water, and finally thrown upon the filter and washed.

This boiling with ammonia will not be sufficient to decompose the subsulphates of iron, which are always produced in liquors containing sulphuric acid in sensible proportions. These precipitates, after washing, should always be redissolved again in hydrochloric acid, before final separation by ammonia, otherwise the results will be far from accurate.

The precipitated and calcined peroxide of iron is anhydrous when it possesses a bright metallic lustre, resembling that of oligist iron. Its powder is of a violet red color. That which is brown and without lustre is not anhydrous, and should be calcined again at a higher temperature. It is well also to try the calcined oxide with the magnet, because it happens sometimes that it has been partly reduced and contains some magnetic oxide.

These observations apply as well to the precipitation of peroxide of iron alone as to that which is mixed with alumina. This latter substance, which, when precipitated alone, is slightly soluble in an excess of ammonia, does not appear to be acted upon by this reagent when a certain amount of iron is precipitated at the same time.

Unless the proportion of alumina is very small, its separation from the oxide of iron by the method indicated in the text, will not be complete in one operation. The remaining oxide of iron ought to be dissolved again in hydrochloric acid, and the treatment by a solution of caustic potassa (or soda) repeated at least once more. The separation is more complete if the powdered mixture of the two oxides be fused with caustic potassa (or soda) in a silver crucible. The oxides are added when the caustic alkali is in a state of tranquil fusion, that is, has ceased to spirt, which is occasioned by the escape of water. The oxides may be calcined, or simply dried at about 100° C. In the absence of a silver crucible, one of platinum will answer, by diluting the caustic alkali with about two or three times its weight of dry carbonate of sodium. There is little, if any, spirting in this case, and the oxides may be immediately mixed with the carbonate of sodium, and a lump of the caustic alkali put on top. The remainder of the operation is explained in the text. This method generally gives an amount of alumina greater than it ought to be, because the caustic alkalies used, soda or potassa, contain themselves a certain quantity of alumina, which increases the weight of the alumina of the ore ; on the other hand, by the method given in the text, the amount of oxide of iron is too great, its weight being increased by a certain quantity of alumina retained. Consequently the weight of the alumina separated ought to be too small, but this is compensated to a certain extent (and sometimes more than it ought to be) by the alumina contained in the solution of caustic alkali employed.

If it be desired to separate the oxide of iron from the alumina without determining the proportion of the latter substance, by itself, this may be effected as follows : enough tartaric acid is added to the solution of the two oxides to prevent any precipitation when an excess of ammonia is added. The iron is then precipitated as sulphide by the hydrosulphide of ammonium, and the alumina is separated by filtration. It is well, when the proportion of alumina is considerably above that of the iron oxide, as in certain clays, for instance, to dissolve the precipitate of iron sulphide upon the filter with hydrochloric acid, and to begin anew the operation, that is, to add tartaric acid, ammonia, and hydrosulphide of ammonium. We have already indicated the manner of washing the precipitate of sulphide of iron, and its further treatment by hydrochloric acid, oxidizing agents, and ammonia.

The alumina may be collected from the filtrate separated from the sulphide of iron, although the operation is very troublesome. On account of the tartaric acid, the solution must be evaporated to dryness, and the residue calcined until white. It may then be dissolved in caustic potassa or soda, or in the acid sulphate of potassium. The volatilization of the ammoniacal salts present, causes a certain loss of alumina, so that this substance had better be determined by difference.

This method is very good for separating small amounts of oxide of iron, which it would be difficult to determine accurately by volumetric analysis. A solution of the two oxides may be divided into two equal volumes : in one, the oxides are precipitated, washed, calcined, and weighed together ; in the other, the oxide of iron alone is determined as we have just seen. The difference of weights gives the alumina.

10*

The calcined mixture of oxide of iron and alumina is very difficult to dissolve in concentrated acids, but is readily acted upon by the acid sulphate of potassium. The writer has in this manner frequently dissolved these oxides in the same state, and the same crucible, in which they were after calcination. Thus pulverization was avoided, and the loss consequent to this operation, because there is always a certain amount of material sticking fast to the agate mortar. It is true that the pulverized portion collected may be weighed again, and the loss ascertained. The action of the fused acid sulphate of potassium, at a low red heat, upon the two oxides is very energetic, and should be made upon a gas burner, the flame of which may be regulated at will and rapidly. The mixture is also frequently stirred with a platinum spatula or wire. When the fusion is complete, the crucible is allowed to cool off, and a new quantity of acid sulphate of potassium is added. By restoring the heat, the mixture is made homogeneous, and we should not wait until all the excess of sulphuric acid has been expelled before we allow the contents to cool off, otherwise the sulphates will be difficult to dissolve.

The exact separation of lime from magnesia is also difficult in one operation. There should be a certain amount of chloride of ammonium in the solution, either added, or formed by the saturation of the acid liquor by ammonia, which is generally the case. The first precipitate of oxalate of calcium is dissolved in diluted hydrochloric acid, and the solution receives a new, but small, quantity of oxalate of ammonium, and is saturated with ammonia. The resulting second precipitate of oxalate of calcium may be considered as being free from magnesia. The filtrate and washings of this second precipitate are added to the former liquors containing magnesia, and should be concentrated if their bulk is too considerable.

As the oxalate of ammonium should be added in excess, and as the fault is generally that of putting in too little of this reagent, it is advisable to add a little more of it to the filtrate separated from the oxalate of calcium, and to ascertain if further precipitation takes place, or not. On the other hand, for the precipitation of the magnesia, too great an excess of phosphate of sodium should be avoided, because there may be produced a precipitate of a double phosphate of sodium and ammonium. But, in order to ascertain if a sufficiency of reagent has been added, a few drops of phosphate of sodium are poured into the filtrate separated from the precipitate of phosphate of ammonium and magnesium. If a new precipitate is observed, we add a little more of phosphate of sodium, collect the precipitate with the former, etc.

There is a case already mentioned in a former note, where the separation of lime from magnesia, cannot be effected directly by the oxalate of ammonium, that is, when the proportion of lime is only about $\frac{1}{100}$ of that of magnesia. The oxalate of ammonium fails to produce a precipitate. It is then necessary to add sulphuric acid to the liquor, which is concentrated if its bulk is too great. The oxalates of lime and magnesia are thus transformed into sulphate of calcium and magnesium, the former of which is insoluble in alcoholic liquors. Alcohol is therefore added, until the solution becomes permanently turbid. The sulphate of lime separated by filtration, contains a certain amount of sulphate of magnesia. It is therefore necessary to redissolve it in water (which is not difficult on account of the small quantity of precipitate), and to precipitate the lime by the oxalate of ammonium.

There is no difficulty in transforming the oxalate of calcium entirely into caustic lime, when the laboratory is provided with a gas blowpipe, with which the crucible and contents may be raised at a white heat. Strongly calcined lime does not absorb water very rapidly, and may be weighed accurately.

Silica, which, before drying, was in the gelatinous state, is generally pure, if the evaporation has been made on a steam-bath. Dried at a somewhat higher temperature, it may have combined with a certain quantity of oxides. It is, therefore, desirable to ascertain the purity of precipitated silica. This is done as follows: The calcined silica should dissolve entirely into a boiling and concentrated solution of carbonate of sodium. By cooling, the silica precipitates again. The operation is best performed in a platinum or silver capsule.

This test should always be applied when the silica separates in flakes or in powder, without passing through the gelatinous state.]

II.—ORES CONTAINING MANGANESE.

Several methods may be employed for analyzing these ores. All those which we shall refer to may be described generally as follows: the ore is first treated with acids, either directly, or after fusion with the alkaline carbonates; the solution, being separated by filtration from the silica, is deprived of precipitable metals, when present, such as copper and arsenic, by treatment with hydrosulphuric acid, then boiled with an excess of hydrochloric acid and a little nitric acid, or, better, with chlorate of potassium; the boiling is continued as long as there is any liberation of chlorine; in this manner all the manganese is changed into a protosalt of manganese, and the iron into the persalt of iron, which is indispensable for the carrying on of the remaining steps. These consist:—

1st. In the treatment of the solution so as to obtain, on the one hand, a precipitate containing the peroxide of iron, alumina, and phosphoric acid, and, on the other, a solution containing the manganese, lime, and magnesia.

2dly. The treatment of the precipitate.

3dly. The separation of the manganese and then the lime from the magnesia, and the determination of these bodies.

Separation of the Peroxide of Iron, Alumina, and Phosphoric Acid from the Manganese, the Lime, and the Magnesia.

First Method: by Ammonia.

The liquid containing the different oxides in solution together with free hydrochloric acid is boiled, and chloride of ammonium is added, and then gradually an excess of ammonia, taking care not to pour in too much at once, that it may not arrest the boiling. The steam and ammonia which are liberated together prevent contact with the air, which must be excluded, in order to avoid the precipitation of the manganese; this, in contact with oxygen, would be deposited as peroxide of manganese. Continue to boil the liquid until the smell of ammonia is gone, replacing, from time to time, the water evaporated with boiling distilled water; then separate rapidly, by filtration, the precipitate formed of the oxides of iron and aluminium, carrying down with them the phosphoric acid; the manganese remains in the solution with the lime and magnesia. The separation is complete in one operation, if the proportion of manganese be small, and if the operation have been carefully performed, otherwise a portion of the manganese remains with the oxide of iron and alumina. In this case, it is necessary to filter the precipitate, to wash it, in order to get rid of the greater part of the soluble matter, and then to dissolve it in hydrochloric acid. Having boiled the solution, we add ammonia; that is to say, we repeat the process employed with the first solution. The precipitate obtained is this time very free from manganese; the solution, when filtered, is added to that obtained at first.

Second Method: by the Alkaline Acetates.

The solution of the ore, when deprived of silica, is neutralized as exactly as possible by carbonate of sodium or ammonium; this has been effected when the liquid assumes a deep red tint. We must be cautious that no precipitate takes place; but, if one should unfortunately form, we must redissolve it by means of a few drops of hydrochloric acid with the aid of heat. Saturation being attained, we add to the liquid thus prepared an excess of the neutral acetate of sodium or ammonium, then boil until the precipitate which is formed collects and deposits readily as soon as ebullition has ceased. The precipitate contains the peroxides of iron and aluminium in the state of basic acetates and phosphates; it is washed in boiling water first by decantation, then on the filter, on which it is placed in order to separate it from the solution, in which the manganese, lime, and magnesia are contained. We make sure that the separation is complete by redissolving the precipitate in hydrochloric acid, and submitting the solution to the same process a second time.

Third Method: by the Alkaline Succinates.

Neutralize accurately the solution of the ore, in the same way as already described; we may even precipitate a little of the peroxide of iron, provided the liquid keeps its deep red color, which shows that the greater portion is still in solution; to the liquid thus neutralized add a perfectly neutral solution of succinate of ammonium or of sodium, and then boil; succinates of iron and aluminium, mixed with phosphates, are deposited. Allow the liquid to settle, then filter, and wash the precipitate in cold water. Exact neutralization is the essential point of this method; to obtain complete separation, it is indispensable to submit the precipitate to a repetition of the process.

The succinates can now usually be obtained in commerce. Where this may not be so, the benzoates answer, that of sodium best, because it can be more easily obtained pure, from benzoic acid of commerce. In the treatment with carbonate of barium it is necessary that the reagents employed be perfectly free from sulphuric acid.

[The succinates are better than the benzoates on account of forming less bulky precipitates than the latter. The calcination of the succinate or of the benzoate of iron, requires to be made in a very oxidizing atmosphere, in order to prevent a reduction by the carbon of the organic acid. The benzoates contain more carbon than the succinates. The greater part of the succinic or benzoic acid may be removed from the precipitate, by treating the latter, after it has been washed as usual with water, by ammonia, which dissolves the organic acid. This operation facilitates the thorough calcination of the peroxide of iron.]

Fourth Method: by Carbonate of Barium.

The solution of the ore is saturated with carbonate of sodium till very little free acid remains; add then an excess of carbonate of barium, stir well, and let it act *cold* for 20 to 30 minutes, taking care to stir the

liquid frequently. Separate the precipitate by filtration. This contains the peroxide of iron, phosphoric
acid, and alumina, combined with the excess of
carbonate employed. Wash with cold water; the
filtered liquid contains, in addition to manganese,
lime and magnesia, a salt of barium produced by the
reaction.

[The protoxide of manganese in a solution as neutral as that
which contains an excess of carbonate of barium, is very readily
oxidized. It is therefore advisable to operate the precipitation in
a corked phial. If the carbonate of barium employed is the
natural *witherite*, or has been calcined, the complete precipitation
may require more than half an hour. Too slight an excess of
carbonate of barium may cause the precipitation of a small proportion of manganese oxide with that of iron.]

Treatment of the Precipitate containing the Peroxide of
Iron, Alumina, and Phosphoric Acid.

If we have operated in accordance with the first
method, we may proceed to calcine the precipitate;
and if we know, by volumetric assay, the weight of
the peroxide of iron, and, by a special experiment,
the weight of the phosphoric acid which it contains,
we can infer the weight of the alumina. We may
also redissolve it on the filter, and separate the oxide
of iron from the alumina, by means of an excess of
potassa, as already indicated in describing the
analysis of ores free of manganese. If, operating by
the second and third methods, we have exclusively
used ammoniacal compounds to obtain neutralization
and precipitation, we may also directly calcine the

precipitate, taking care to allow a current of air in order to avoid the reducing action of the organic acids. But if neutralization has been effected by means of carbonate of sodium or of potassium, or again, if we have made use of the acetate or succinate of sodium, it is to be apprehended that the oxides have retained alkaline salts, however carefully the washing may have been performed. In this case, we must redissolve the still moist precipitate in the least possible amount of hydrochloric acid, and precipitate it again with ammonia, or else treat the solution with an excess of potassa, in order to separate the alumina from the oxide of iron.

So far as the writer's experience goes, the separation of alumina and iron by caustic potassa is troublesome and not satisfactory. Wöhler's method is greatly preferable, which is this: Neutralize the dilute solution of both bases with carbonate of sodium, add to it hyposulphite of sodium, and boil until sulphurous acid ceases to be evolved. The alumina collects as a pretty dense precipitate, which only needs to be washed, calcined, and weighed.

To the solution, after concentration, add chlorate of potassium and hydrochloric acid, filter to separate free sulphur, and then precipitate the iron by ammonia. Parnell has described another method, but probably not as good as the above.

Iron may be also separated from alumina as a volatile chloride by just passing hydrogen over the oxides, heated bright red (in a boat), in a porcelain tube, and then passing dry chlorine over them. For every-day analysis this method (even if absolutely exact, which we do not know that it is) may prove too tedious, however.

The carbonate of barium employed in the fourth method renders the operation a little longer. The precipitate is redissolved on the filter by means of dilute hydrochloric acid; the filter is then washed with warm water, slightly acidulated with the same

11

acid, and the filtrate is diluted; after boiling, the baryta is precipitated with dilute sulphuric acid. Let it stand, separate the sulphate of barium by filtration, and then precipitate the oxides of iron and alumina and the phosphoric acid they bring with them, by the aid of ammonia. Filter, wash the precipitate, and calcine. If the ore were sulphurous, with the oxides and the carbonate of barium of the first precipitate, there would also be a small quantity of the sulphate of barium, which would not, however, interfere with the rest of the operation.

This is not quite the writer's experience. If sulphate of barium be formed in any way, whether by sulphuric acid derived from oxidized sulphur in the ore, or from impurity in the reagents, it appears to take up a minute quantity of iron, from which it cannot be freed unless after decomposition.

[The washed and calcined sulphate of barium is treated with hot and concentrated hydrochloric acid, which will eliminate the iron and some other impurities. The solution is then diluted with water, two or three drops of sulphuric acid are added, and the sulphate of barium is separated by filtration. The addition of a few drops of sulphuric acid is intended to precipitate the chloride of barium formed by the action of hydrochloric acid upon the sulphide of barium often found in the sulphate, and resulting from the partial reduction of the latter, during calcination, by loose fibres of the filter, or some other cause.]

Treatment of the Liquid containing Manganese, Lime, and Magnesia.

If we have used carbonate of barium for the precipitation of the iron, the liquid will also contain a salt of barium; in this case, before submitting the liquid to either of the following processes, we must get rid of the baryta by means of dilute sulphuric acid.

First Method: by Hydrosulphide of Ammonium.

The clear, colorless liquid is poured into a conical (or silver assayer's) matrass and heated slightly; add chloride of ammonium, if it do not already contain that, in consequence of the employment of ammonia or a salt of ammonium for the precipitation of the iron; then precipitate the manganese with the hydrosulphide of ammonium. The addition of chloride of ammonium is intended to assist in the precipitation of the sulphuret of manganese. If the separation of the iron has been well effected, the precipitate should be of a very clear rose-color; the least trace of iron left in the solution is attended with a gray tint, more or less deep, or even with blackness, if the proportion of iron be considerable. After precipitation, close the matrass with a cork or a watch-glass, and let it stand for twelve hours in a warm place.* It happens sometimes that during this period the precipitate assumes a greenish tint; this change may be accounted for by the fact that the sulphuret of manganese, which is hydrated at the moment of precipitation, loses some of its water of hydration after a more or less extended period.

The filtration and washing of the precipitate must

* This prolonged repose is necessary for the complete precipitation of the manganese.

be effected with certain precautions. Filter off the clear liquid from the precipitate, and add the latter to an aqueous solution of chloride of ammonium, to which some drops of the hydrosulphide have been added; let it settle for a while, decant again upon the same filter, and add the filtrate to the solution first decanted; repeat the washings several times, then, having collected all the precipitate on the filter, wash it with distilled water containing a little of the hydrosulphide and chloride of ammonium; diminish the amount of the latter as the washing proceeds, and towards the end of the washing suppress it entirely. We must take care that during the washing the precipitate be not too long in contact with the atmospheric air, in order to avoid a small quantity of the sulphuret of manganese becoming oxidized and changing into the protosulphate, which, being soluble, would be carried over in the filtration. The addition of the hydrosulphide of ammonium is to prevent this oxidation. Nevertheless, we should take care to keep the filter constantly nearly full, and to cover the funnel with a glass plate. The funnel must be large enough, and have a ground edge.

The first of the filtered liquid is generally somewhat turbid; in this case, we should filter until the liquid flows clear, and then change the jar which receives it, and throw back the turbid portion upon the filter. The washing being ended, dry the precipitate, and calcine it with the precautions indicated already, or else treat the precipitate and the filter with dilute hydrochloric acid, avoiding an excess;

the sulphuret of manganese is changed into the soluble chloride, and hydrosulphuric acid liberated. Heat until the latter has been completely driven off, and filter to get rid of the matter (sulphur and fibres of paper) in suspension. Add to the filtered and boiling liquid some carbonate of sodium, in very small quantities at a time, to avoid effervescence arising from the liberation of the carbonic anhydride, until, by means of strips of reddened litmus paper, we ascertain that the liquid has acquired an alkaline reaction; then leave the precipitate of protocarbonate of manganese to settle. After two or three washings by decantation with hot water, throw it on a filter and continue the washing; this effected, dry the carbonate, calcine it strongly in contact with the air, and then weigh. By calcination the carbonate or sulphide, by the action of the oxygen of the air, is converted into the intermediate oxide of manganese (*manganoso-manganique*).

[The transformation of the sulphide of manganese into manganoso-manganic oxide, by direct calcination, is made only when the quantity of sulphide is small.]

The liquid deprived of the manganese sulphide being heated to boiling, pour in nitric or hydrochloric acid in small proportions at a time until the complete decomposition of the hydrosulphide of ammonium (which we ascertain by the smell), filter the liquid to separate the sulphur deposited, neutralize with a little ammonia, separate the lime and magnesia according to the method indicated already, precipitate the first with oxalate of ammonium, and the second with phosphate of sodium, in the presence of

11*

a large proportion of free ammonia; the liquid contains, as a result of the process, the necessary amount of chloride of ammonium.

Second Method: with Chlorine Gas.

To the clear, neutral, and sufficiently diluted solution containing manganese, lime, and magnesia, add (if there be not already sufficient present) some acetate of sodium, raise to a temperature of from 50° to 60° Cent., then pass through it to saturation a current of washed chlorine gas. The manganese is partly precipitated in the state of hydrated peroxide; when the liquid is saturated with chlorine, add an excess of ammonia to insure the complete precipitation of the manganese; boil the liquid until the free ammonia has completely disappeared. Filter and wash the precipitate, and as it retains akaline salts, which during the calcination would occasion the transformation of some of the oxide of manganese into manganic acid, redissolve it in boiling hydrochloric acid, and precipitate the liquid thus obtained with carbonate of sodium, as indicated in the preceding method.* The lime and magnesia remaining in the liquid are treated in the usual way.

Rose does not recommend this method, if, as is most generally the case, the solution contain salts of ammonium; he apprehends the formation of chloride of nitrogen, a most formidable explosive

* It is always an advantage in operating with manganese to make sure of its complete precipitation by means of the hydrosulphide of ammonium.

compound. We have never observed this, and are of opinion that chloride of nitrogen, being a combination which decomposes easily in contact with a number of foreign bodies, is not likely to remain if it be formed in the present instance. Instead of a current of chlorine gas, we may employ water saturated with chlorine.

Third Method: with Bromine.

Properly speaking, this is only a modification of the preceding method: the cold liquid separated from the peroxides of iron, etc., is rendered very alkaline with ammonia; some drops of bromine are added, or, still better, a certain quantity of bromine water, and it is left to rest for twenty-four hours in a corked matrass. After this time, if the liquid contain free bromine, add ammonia, heat and filter; wash the precipitate, and if there be no salt with alkaline base present, we may directly calcine the peroxide of manganese, in order to transform it into the intermediate oxide (*manganoso-manganique*); if alkaline salts be present, we must redissolve it with hot hydrochloric acid, and precipitate it with the usual precautions, with carbonate of sodium.

The carbonate of manganese thus obtained does not retain the alkalies with as much energy as the peroxide; we must take care to employ the carbonate of sodium in as slight an excess as possible.

The same method as already given will serve for the separation of lime and magnesia.

[The precipitation of the manganese may also be effected as follows: To the neutral filtrate separated from the precipitate of iron oxide and alumina by the treatment with acetate of sodium, add a few drops of acetic acid and bromine, and boil until the bromine smell disappears.]

Remarks.

1st. It may happen that we only desire to ascertain the amount of manganese in an ore.

In this case it is sufficient to treat the liquid obtained by the action of the acids upon the ore with acetate of sodium, without regarding the insoluble matter it may contain. After the reaction, conducted in the manner already described, filter, and in the liquid thus obtained, deprived of the peroxides of iron and aluminium, precipitate the manganese in accordance with any one of the methods previously described.

2d. When the ore contains zinc the latter is precipitated with the manganese, if we operate in accordance with the first method; in this case we must subtract from the weight of the manganoso-manganic oxide thus obtained the weight of the oxide of zinc. This is to be found by a special assay. In the second and third methods the zinc remains in the solution with the lime and magnesia. Before precipitating these two bases, we get rid of the excess of chlorine or bromine by boiling, and separate the oxide of zinc by the hydrosulphide of ammonium, following precisely the process before indicated for the precipitation of the manganese.

3d. We can determine the amount of manganese

very rapidly if the ore contain neither lime nor magnesia, or contain so little that we may disregard them; in this case, we treat the liquid deprived of the iron and alumina by means of the acetate of sodium, directly with the carbonate of sodium.

III.—DETERMINATION OF THE LOSS BY CALCINATION.

The loss of weight that an ore sustains by calcination is due to the evaporation of water, the liberation of carbonic anhydride, in consequence of the decomposition of carbonates, or to the combustion of organic matter. The presence in the ore of the protoxides of iron and of manganese, or of the sulphides, prevents our considering the loss by calcination as exactly expressing the proportion of water, organic matter, and carbonic anhydride united; in fact the protoxide of iron, passing by calcination in contact with the air, into peroxide, compensates by the increase of weight thus induced for a part of the loss sustained by the ore during the operation; the sulphides, on the other hand, passing into oxides increase the loss; the presence of manganese also falsifies the calculation of the loss by calcination.

The operation is effected as follows: Weigh, in a small weighed capsule with a flat bottom, about two grammes of very finely powdered ore, then heat the capsule gradually to bright redness in a muffle; after some time withdraw, cool it in the drying bell, and weigh. To make sure that the calcination be complete, heat again for a quarter of an hour, and repeat the weighing. If necessary, repeat the process several times, until the two last weighings agree. Generally speaking, it is sufficient to know the total loss produced by calcination, but it may be of importance to know the actual quantity of water and carbonic anhydride comprised in this loss. If so we must have recourse to special assays, which we shall describe.

IV.—ASSAY FOR WATER.

An ore may contain water, either absorbed hygroscopically or in chemical combination. We must ascertain separately the quantity of water in each of these conditions.

Determination of Water not in Combination.

When an ore is too moist to be directly pulverized, the amount of hygroscopic water is ascertained by two successive operations. Take one or two kilogrammes of the ore grossly powdered with the hammer, spread it on an iron dish or plate with raised edges in a warm place, the temperature of which we can regulate by the thermometer, and which should not exceed 100° Cent. Stir the mass occasionally with a spatula, and when deemed sufficiently dry weigh again, and then pulverize the ore finely for further experiment. The ore still contains some hygroscopic water; to ascertain the amount, weigh two or three grammes of pulverized ore in a platinum capsule with a flat bottom, and subject them for a sufficient time to the temperature of 100° Cent. in an oven where the air is constantly renewed. Then cool in the drying bell, and weigh again.   To be certain of the accuracy of the result—that is to say, of the complete driving off of the water—it is necessary to repeat the process till two consecutive weighings agree. In many industrial laboratories, the ore is dried until capable of being pulverized, and no account is taken of the loss it sustains by this treatment; it is then submitted to volumetric assay, and the figure thus obtained is considered to represent the real amount of iron. It is obvious that experiments made under these conditions are deficient in accuracy in two respects. We do not know the contents of the ore as it is, nor its contents after desiccation at 100°. This last must be viewed as the truest, for it is the only state which remains constant, and upon which the further processes to which the ore is subjected have no influence.

The quantitative determination of the iron supposed to be contained in an ore is made with a portion of the substance more or less dried; the richness of the damp ore is calculated by the iron found in the dried ore, and on the proportion of water it lost by desiccation. For instance, an ore containing 22 per cent. of hygroscopic water, and in its dry state 34 per cent. of iron, when damp would only contain $\frac{(100-22)\,34}{100}$ per cent.—that is, 26.32 per cent. of this metal.

Determination of Water in Chemical Combination.

It is evident that if the ore contain neither carbonic acid nor organic matter, nor sulphides, nor protoxide of iron, nor oxide of manganese, the proportion of water in combination can be very easily obtained by subtracting from the loss which the ore sustains by calcination the amount of hygroscopic water found in it by simple desiccation. This is, however, an exceptional case, and we are generally obliged to have recourse to direct weighing in order to obtain the amount of water of combination.

The general process is as follows: The ore is placed in a tube of refractory German glass, and heated in a current of dry atmospheric air; the water evaporated is carried off and retained in a tube of dry chloride of calcium, accurately weighed beforehand. By the increase in weight of the absorbing apparatus, we learn the weight of the water. In order to obtain accurate results, the apparatus (Fig. 9), and the

Fig. 9.

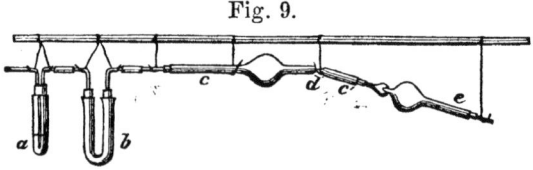

accompanying method of operating, are, in our opinion, the best: *a* is a small tube containing

sulphuric acid, and intended to control the rapidity
with which the air traverses the apparatus; *b* is a
tube, U-shaped, filled with chloride of calcium; it
serves to dry the air completely before that reaches
the ore. The ore is contained in the ball of the tube
cc', made of refractory glass, narrowed and bent in *d*,
as the figure indicates; *e* is a tube containing chloride
of calcium in small fragments. A current of air is
sent through the apparatus by placing it in communication at *e* with an aspirator, consisting of a large
bottle full of water, which is emptied by a siphon.
The aspirator bottle is similar to that often used for
keeping distilled water (see Fig. 1). A small tube
filled with chloride of calcium is adapted to this
aspirator to prevent any vapor from the water entering tube *e*; a screw clip-cock regulates the flow of
the water. We operate as follows: In a glass tube
closed at one end, and smaller than tube *c*, weigh
from two to five grammes of the ore; then introduce
the open part of this tube into tube *c*, and by shaking
slightly while turning the tube between the fingers,
cause the ore to fall into the ball, taking care that it
does not go beyond it. Weigh tube *e*, furnished with
a cork that serves to unite it to *c*, then join the
different parts of the apparatus by means of corks
and caoutchouc tubes; bring the aspirator into play,
and heat the ore gradually and slowly with a spirit
lamp or Bunsen gas flame. Move the lamp gradually
as far as *d* in order to expel the water beyond the
narrow part, then separate *c'* from *c* either by a stroke
of a file at *d*, or by fusing to a neck the glass at the
narrow part by the blowpipe, and weigh *c'* and *c*

12

together ; then separate the two, wash and dry c', and weigh it in its turn. By means of these data we easily determine by the difference the weight of the water driven off from the ore. If the ore has been previously dried at 100°, this weight is the weight of the water in combination; in the opposite case we should allow for the proportion of hygroscopic water.

V.—ASSAY FOR CARBONIC ANHYDRIDE.

Carbonic anhydride is found in ores, in combination with different oxides. The principle of the method usually employed to ascertain its proportion is as follows: Put a known quantity of the ore to be tested in one part of a small glass apparatus, sufficiently small and light to be placed on the balance pan, and consisting of two separate vessels capable of communicating with each other. Place the weighed ore in one vessel, and pour some strong acid into the other, and unite the two; the carbonic anhydride is expelled, and if care has been taken to prevent the escape of any vapor with it, and to secure its complete expulsion from the apparatus at the conclusion of the operation by drawing through a current of air, we are enabled to ascertain its amount by the difference in weight. This very simple method gives sufficiently accurate results. We cannot give the details of all the apparatus proposed to be used for this purpose. Mohr's apparatus, as modified by Bunsen, is, by construction, easily handled, and very suitable.

This apparatus (Fig. 10) consists of a small glass flask, furnished with an S-shaped tube, with a ball in-

Fig. 10.

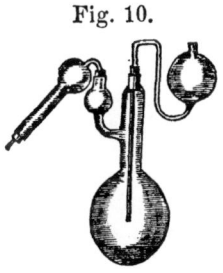

tended to contain the acid required for the decomposition of the carbonates, and a small tube for chloride of calcium, bent downwards to make the apparatus more compact, and render it easier to weigh. Put a weighed quantity of the ore into the flask with a little water; the less carbonic anhydride contained in the ore, the more ore we should employ. Affix the drying tube and the S tube, having filled the ball of the latter with ordinary nitric acid. Then dry the apparatus externally, carefully with a dry cloth, and weigh it. By drawing (with the mouth) the air from the tube containing the chloride, we cause a little of the acid to enter the flask; after effervescence has abated, cause more acid to enter, and continue this until no further effervescence be produced by the addition of the acid; then heat gently, and suck with the mouth again through the tube containing the chloride, in order to cause a small current of air to pass into the flask, and expel

from it the least trace of carbonic anhydride. Cool
the apparatus, dry it, and weigh after an hour.

VI.—ASSAY FOR THE PROTOXIDE OF IRON.

The simplest way to make this assay is to employ
the volumetric method by permanganate of potassium
as in the case of ores. All oxidizing action should
be avoided during the process of solution, which is
to be effected by means of hydrochloric acid only
(care being taken not to employ too great an excess)
in a flask through which a current of carbonic
anhydride is passed to prevent access of the atmosphere.

A more practical method, especially if · one has
several assays to make together, is to put into the
flask with the ore a small portion of crystallized
carbonate of potassium or of sodium; under the
influence of the acid, the carbonate will cause the
liberation of carbonic gas, which will rapidly expel
the air from the flask. This should be furnished
with the stopper and valve before described. The
solution being obtained, add some sulphuric acid to
expel the greater portion of the hydrochloric acid;
heat to assist its liberation. Dilute with cold water,
and transform the protosalt into the persalt by means
of the standard solution of permanganate, the volume
of which employed will then indicate the quantity
of the protoxide contained in the ore. We might

also determine in addition the amount of the peroxide of iron by means of chloride of tin, and knowing the total content of iron of the substance, infer the quantity of the protoxide. The results furnished by the latter method have the advantage of not being influenced by the presence of hydrochloric acid, and thus the use of sulphuric acid to expel it is unnecessary.

[This method, to be practicable, requires that the ores should be easily dissolved in hydrochloric acid, and should contain no oxide of manganese of a higher degree of oxidation than the monoxide. Manganiferous iron ores generally contain the manganese at a higher degree of oxidation, and consequently by the treatment with acids, the protoxide of iron is peroxidized, and the analysis is of no value.

If the ores are difficult to decompose with hydrochloric acid, it is difficult to effect their solution entirely without the contact of the air.

Ores, or products of art insoluble in acids, may be decomposed by treating them with hydrofluoric acid, in an apparatus through which steam and carbonic anhydride are passing. They may also be fused with powdered glass of borax in a closed platinum crucible, held itself within one or two other crucibles filled with calcined magnesia, in order to avoid the contact of the air. The fused product, or an aliquot part of it, is then dissolved in hydrochloric acid within a vessel filled with carbonic anhydride. These operations require a great deal of care, and are more adapted for scientific determinations of compositions of minerals than for practical every-day work.

There are other methods by which the perchloride of iron is determined by means of metallic copper or silver, and the protochloride is ascertained by difference, after the whole amount of iron in the solution is known. For these we refer the reader to larger analytical works.]

VII.—ASSAY FOR PHOSPHORIC ACID.

REMARK.—Arsenic behaving generally like phosphorus, its presence in the ore requires a slight addition to the different methods about to be described: the liquid when deprived of silica must be boiled after the addition of the sulphurous anhydride or sulphite of sodium to reduce the arsenic acid to the state of arsenious acid. When the boiling has been continued until the smell of sulphurous anhydride has completely gone off, precipitate the arsenic as a sulphide, by means of a current of hydrosulphuric acid, filter, and in the filtrate solution ascertain the quantity of phosphoric acid by one or other of the methods now to be described. The reduction of the arsenic acid and the persalts of iron by sulphurous anhydride or sulphite of sodium, has the advantage of being more rapid than that effected by hydrosulphuric acid; the production of a large quantity of free sulphur which would go down with the precipitate of sulphide of arsenic is thus also avoided.

First Method: by the Sulphite of Sodium.

Add to the solution of the ore deprived of silica some sulphite of sodium, in order to bring back the persalt of iron into the state of a protosalt. When the reduction is nearly finished (which we recognize from the loss of color in the liquid), boil, to obtain the complete expulsion of the sulphurous anhydride whether free or in combination; for this purpose we must use an excess of hydrochloric acid, which also prevents the precipitation of basic salts in consequence of the ebullition; as soon as the liquid boils drop in some solution of the perchloride of iron in such proportion as will enable the oxide of iron, of which it contains the elements, to supersaturate that and the phosphoric acid contained in the ore, being cautious as to a too great excess. We know that the amount of the perchloride is sufficient, when the precipitate assumes a decidedly red-brown color, a color which the phosphate of iron does not possess, and which arises solely from an excess of the oxide.

The surest plan is to make use of a standardized solution of the perchloride of iron, containing 15 milligrammes of iron to each cubic centimetre, and to employ of this solution as many cubic centimetres as shall result from the multiplication of the supposed quantity of phosphoric acid present expressed in hundredths, by the number of grammes of ore employed.

The precipitate mentioned above contains all the phosphoric acid united with the oxide of iron and alumina, but exempt from lime, magnesia, and manganese; exempt, also, from too great an excess of the oxide of iron, this having remained in solution in the condition of a protosalt of iron. Filter while hot; the liquid passes clear at first, but is soon discolored by the oxidation of the salt of iron, which is, however, of no importance.

The precipitate collected on the filter is dissolved by means of hydrochloric acid: to its solution add tartaric acid, chloride of ammonium, and a great

excess of ammonia; the latter will produce no precipitate if care has been taken to employ a sufficient
quantity of tartaric acid. Finally, by the addition
of sulphate of magnesium we precipitate the phosphoric acid in the state of a double phosphate of
ammonium and magnesium. Stir the fluids with
a glass rod, avoiding contact with the sides of the
vessel for the reasons given already in reference to
the quantitative analysis of magnesia.

It is scarcely possible to avoid stirring a solution in a precipitating jar without touching its sides more or less with the glass
rod. It, moreover, may be remarked that where the amount of
magnesia present is small, or under some circumstances is
reluctant to precipitate, this may be much promoted (as Wollaston
long ago pointed out) by drawing a glass rod over the inner
surface of the glass vessel.

If the glass surface be kept quite free from actual scratches, the
adherence of some of the minute crystals of the double phosphate
to the points touched by the glass rod is really of little importance.
They can generally be detached by water and shaking; or if not,
by gentle rubbing with the end of a bit of clean wood, which is
carefully washed afterwards.

The precipitate filtered after twelve hours' rest,
and washed with ammonia water, still retains a small
quantity of foreign bases; to get rid of these, we
must redissolve it in hydrochloric acid, add to this
solution a little tartaric acid, and precipitate again
with ammonia. Filter, wash, and weigh after careful calcination.

According to Frésénius, the second solution of the
double phosphate, followed by a second precipitation
with ammonia, is more injurious than useful, as
involving a loss in the phosphate of magnesium, of
which the quantity is never very great.

[The experience of the writer, and we believe, of other chemists, is that the presence of tartaric acid sometimes prevents the precipitation of the phosphoric acid.

Citric acid has been proposed as a substitute, and Frésénius gives the following method, applicable to iron ores and to the analysis of the metal itself. The hydrochloric solution is deoxidized by the sulphurous anhydride or the sulphite of sodium, and then boiled until all sulphurous smell is gone. When cold, the liquor is neutralized by ammonia, acetate of ammonium or sodium is added, and the whole is brought to a boil.

If no red precipitate takes place, a few drops of chlorine or bromine water or of perchloride of iron, are added, until we obtain a red precipitate, which contains all the phosphoric acid combined with alumina and peroxide of iron.

This precipitate is immediately collected upon a filter, while the liquid is hot, afterwards dissolved in hydrochloric acid, and citric acid, then ammonia, and sulphide of ammonium are added to the middling warm solution. A precipitate of sulphide of iron is formed, which collects more easily in a warm liquor than in a cold one, especially when organic substances such as tartaric or citric acid are present.

The sulphide of iron is collected upon a filter, and the filtrate (holding alumina and phosphoric acid) and washings are boiled down, with contact of the air, in order to separate the sulphur. After filtration, the phosphoric acid is precipitated with the ordinary magnesia mixture.]

Before proceeding to assay either iron ore or metal for phosphorus (or indeed for any other possibly present or absent body), the existence of the particular body should be insured by some qualitative testing, the methods for which are to be found in the various standard works on Analysis, and need not be here repeated. As, however, the presence or absence of phosphorus in iron or its ores is a matter of high metallurgic import, such qualitative examinations have to be often repeated, and with care. It seems worth mention that M. Salet states that phosphorus in either steel or iron may be recognized by spectroscopic examination of the hydrogen given off by their solution in hydrochloric acid. This method may be applied to iron ores by bringing a small portion to the state of metal by ignition in a current of dry hydrogen, or by direct fusion.

Second Method: by fusion with the Alkaline Carbonates.

The ore is fused with four to six times its weight of sodic carbonate of potassium; the fused mass is treated with hot water until its complete disintegration; by these operations the phosphoric acid which has combined with the alkaline bases passes into the liquid and leaves in an insoluble state the oxides of iron, the alumina, and the oxides of the other bases which existed in the ore. These matters being separated by filtration, the filtered liquid is neutralized by means of hydrochloric acid, then evaporated to dryness in order to obtain the silica in an insoluble form; the residuum is treated with water, which redissolves everything except the silica. Filter, and in the filtrate previously mixed with ammonia and chloride of ammonium, precipitate the phosphoric acid by means of the sulphate of magnesium.

The operation is completed in the manner already detailed.

[If the proportion of silica in the ore be small, and that of alumina comparatively great, the alkaline solution may contain alumina, which will remain in the liquors and will be precipitated alongside with the phosphoric acid by the above process. In such case, it is necessary to add to the ore a certain quantity of finely divided silica (that kept from former analyses will answer well), which, during the treatment with the sodic carbonate of potassium, will form a double silicate of aluminium and sodium (or potassium) insoluble in water. The alumina may be prevented from precipitating by adding citric or tartaric acid to the hydrochloric solution, after its separation from the silica.]

Third Method: by Tin.

The ore is dissolved with nitric acid only ; dilute the solution with water, and filter to separate the silica; add concentrated nitric acid to the filtered solution, then put into it a known weight of laminated tin, the weight of which should represent 8 to 10 times the probable weight of phosphoric acid contained in the ore; heat until all the tin be oxidized and the liquid clears easily when left at rest, by letting the oxide deposit. The tin, completely reduced to oxide, carries with it the phosphoric acid in combination; both these compounds, the oxide and the phosphate of tin, are insoluble in nitric acid.

The precipitate is thrown on a filter, washed, and calcined, carefully avoiding all reducing action; it is then weighed, and by subtracting from its weight the weight of the peroxide of tin corresponding to the tin which has been dissolved from the laminated tin employed,* we obtain the weight of the phosphoric acid.

The solution of the ore should be quite free from hydrochloric and sulphuric acids.

This method being in some respects doubtful, it is better to modify it in the following manner: add to the nitric solution a sufficient quantity of tin, the weight of which is not necessarily known. When

---

* This weight is determined either by calculation (one part of tin gives 1.2712 part of anhydrous oxide) or, if preferred, by a previous experiment, tin not being in general absolutely pure.

the peroxide of tin which forms is well deposited,
wash it several times by decantation, always pouring
the decanted wash liquid through a filter. After
sufficient washing, place the oxide in a platinum
capsule, dissolve it by a moderate heat with a small
quantity of a concentrated solution of caustic potassa.
Dissolve also with a little caustic potassa the peroxide adhering to the filter, and unite the liquids;
they should be limpid; saturate with a current of
hydrosulphuric acid; in this manner the oxide is
changed into a sulphide, which, owing to the sulphide
of potassium simultaneously formed, remains in
solution. Precipitate the sulphide of tin, by adding
acetic acid or dilute sulphuric acid until a slight
acid reaction is found; allow the precipitate to settle
and then filter. The filtrate contains phosphoric
acid in the state of phosphate of potassium; precipitate the phosphoric acid by sulphate of magnesium
in the presence of chloride of ammonium, and an
excess of ammonia, as before directed.

Fourth Method: by the Molybdate of Ammonium.

The solution obtained by the action of hydrochloric
or nitric acid on the ore, when separated from the
silica, is concentrated, if necessary, by evaporation to
a volume of about 10 c. c., then poured into an excess
of molybdic solution, prepared as indicated under
that reagent (Part I.). This solution should be in
such a proportion as to contain at least thirty times
as much molybdic anhydride as there is phosphoric
acid in the solution of the ore. Stir the liquid with-

out touching the sides of the vessel, then let it stand at a temperature of from 30° to 40° Cent., from twelve to twenty-four hours. A yellow precipitate forms, which is separated by filtration, washed with molybdic solution, diluted with its own volume of water, and redissolved afterwards on the filter with the aid of ammonia. Pour into this ammoniacal solution a solution of sulphate of magnesium, which precipitates the phosphoric acid in the ordinary form of double phosphate of ammonium and magnesium.

The precipitation of phosphoric acid by molybdate of ammonium is not unfrequently uncertain in result if certain conditions not sufficiently referred to in the text be not attended to.

An aqueous solution of molybdate of ammonium is to be preferred to that in nitric acid commonly employed. Its strength is but from 50 to 60 grammes to the litre of water. Phosphorus is not precipitated by this reagent from neutral solutions, and on the other hand *strongly* acid solutions retard or even resist precipitation.

The following are M. J. Parry's instructions for the precipitation, to the accuracy and value of which the writer can testify: Add ammonia to the solution until complete precipitation of peroxide of iron. Add cautiously as much nitric acid as is just sufficient to redissolve the precipitated peroxide. Bring the solution to boil, and add the molybdate of ammonium in the proportion of about 30 cubic centimetres to the $\frac{1}{4}$ litre of iron solution, which should be contained in a flask (or a little more, if it be very rich in phosphorus). The usual yellow crystalline precipitate may appear at once, but if not, boil briskly again for a few minutes, add a very few drops of nitric acid, and shake the closed flask vigorously at intervals, and continue to add a drop or two more of nitric acid, until a distinct precipitate is observed to commence. The ebullition must now be stopped, or a bulky flocculent precipitate will rapidly form; but the flask should be kept hot and as near to the boiling point as possible (without actually boiling), and shaken briskly now and then. In from an hour or two to four or five hours the whole of the phosphorus

13

will usually have precipitated in a good granular form. If these details be fully observed it is seldom necessary to repeat the process in order to obtain the whole of the phosphorus present.

[Some persons determine the amount of phosphorus from the weight of molybdo-phosphate of magnesium obtained. This is not accurate, as the composition of the precipitate varies with the mode of operation. The precipitate should be redissolved in ammonia, and the phosphorus determined in the ordinary form of double phosphate of ammonium and magnesium.]

Remark.

Instead of ascertaining the amount of phosphoric acid directly from the ore, it is sometimes preferable to determine the phosphorus in the cast-iron obtained by the dry method. The reason generally why we wish to know the quantity of phosphoric acid contained in an ore is the influence that its presence exerts on the quality of the pig and bar iron produced from the ore. Now as all the phosphorus of the ore does not pass into the pig-iron made from it, we are disposed to think that the last-mentioned method gives results of a more practical character than those obtained by the direct assay of the ore. We refer those who share this opinion to the subsequent article—Determination of Phosphorus in Cast-iron.

We must, however, remark that the proportion of phosphorus found in cast-iron, obtained by crucible operations, is generally much less than that found in pig-iron produced by the treatment of the same ore in the blast-furnace—a consideration which is in favor of the direct determination of the phosphoric acid in the ore itself.

VIII.—ASSAY FOR SULPHUR.

Sulphur is usually found in the ore in the state of sulphide, but occasionally it is also found—at least partially—as sulphate of calcium or of barium. To ascertain the proportion of sulphur in the condition of a sulphide, we must by oxidation change it into sulphuric acid. Several methods have been proposed for effecting this oxidation; we shall mention the two following.

First Method: by fuming Nitric Acid.

Finely pulverized ore is treated hot with a very large quantity of fuming nitric acid, which should be quite free from sulphuric acid; and this it is well to ascertain beforehand. When the action has apparently ended, add a little hydrochloric acid, and continue to heat at about 100° Cent. for some time longer. The solution is then evaporated to dryness with a new addition of hydrochloric acid, so as to expel the greater portion of the nitric acid. The residuum, being first moistened with concentrated hydrochloric acid, is treated with hot water, and finally filtered in order to separate the insoluble matters* in the liquid, heated to about 100°, and very much diluted; we precipitate the sulphuric acid

---

* To insure the complete oxidation of the sulphur, it is well to treat the insoluble residuum with more nitric acid. When the ore is not completely attacked by the treatment indicated, it is better to have recourse to the second method here described.

by means of chloride of barium. Too large an excess
of this salt is to be avoided, as its presence hinders
the complete washing of the precipitate. The precipitate being deposited and the liquid cooled, pour
the latter only on a filter, and wash the precipitate
several times by decantation with water, acidified
slightly with hydrochloric acid; then place it on the
filter, and wash it again there with hot water, until
the washings are no longer affected by the addition
of a drop of dilute sulphuric acid. Dry the filter
with its contents, and calcine carefully. After calcination, the precipitate should be perfectly white and
powdery; if it be red and agglutinated, the washing
has been insufficient; in this case, it must be purified
in the following manner: Drop upon the precipitate
in the capsule in which it was calcined some drops
of hydrochloric acid. Stir it well, and break it up
by means of a glass rod, then pour it into a vessel
containing a small quantity of hot water, and boil
it; when the sulphate of barium has been deposited,
pour the clear liquid on to a filter, add more water,
slightly acidified with hydrochloric acid, boil again,
and filter as soon as the precipitate has been deposited;
then treat it as described above.

The nitric acid may be replaced by aqua regia, but
the first is generally preferred, as less likely to cause
a loss of sulphur in the state of gaseous combination,
or as vapor of sulphur.

[It is sometimes difficult to obtain fuming nitric acid, free from
sulphuric acid. The oxidation of the sulphur in the ore may be
effected by a mixture of chlorate of potassium with hydrochloric
acid or ordinary, but pure, nitric acid. To prevent loss by spurt-

[ing, the ore, mixed with about twice its weight of chlorate of potassium, is placed in a porcelain capsule covered with an inverted funnel. The acid is then added by small quantities at a time.]

Second Method: by fusion with the Alkaline Carbonates and Saltpetre.

This method is based on the transformation of sulphur into sulphuric acid in the dry way.

The matter to be assayed is well mixed with four times its weight of sodic carbonate of potassium, and twice its weight of nitrate of potassium, all being dry and well pulverized. Throw this mixture in small portions at a time into a platinum crucible, heated to redness; at every addition, allow the reaction to take effect, and heat the crucible again, taking care to keep it closed, to prevent spurting over. After the last addition heat till the mass is in tranquil fusion; then let it cool. The fused mass is treated with dilute hydrochloric acid until its complete disintegration, and the solution is then evaporated to dryness in order to render the silica insoluble, and to decompose the nitrates; the residuum, being first moistened with hydrochloric acid, is then treated with hot water; a solution is thus obtained which, after the separation of the silica by filtration, is perfectly limpid, and in which we ascertain the amount of sulphuric acid by chloride of barium, in the manner already indicated. The washing of the sulphate of barium must be even more carefully performed than in the preceding case, for the liquid contains a large amount of fixed matter, which the sulphate of

13*

barium has a tendency to retain mechanically. This method is especially applicable when the ore contains a large proportion of sulphur, as, for example, iron pyrites.

Porcelain or thin and smooth Cornish assay crucibles answer quite well for this oxidation of sulphur by fusion, if the heat be adroitly managed so that too rapid changes of temperature shall not crack the crucible. Platinum crucibles are no doubt highly convenient, but they are rapidly made rough and porous, and destroyed by this use of them; and gold crucibles are luxuries not for industrial laboratories.

Few analysts are aware that in determinations of sulphur by oxidation in this way, it is absolutely necessary to avoid the use of coal gas flames, and best even contact with coal or coke fuel. The quantity of sulphur (chiefly as bisulphide of carbon) with which all British, and nearly all the coal gas of commerce is loaded, is so great that it is not possible to fuse nitre over a Bunsen or other flame of such gas without being able to detect appreciable quantities of sulphate of potassium in it. In fact there is no better test for sulphur in this state in coal gas than to dip the exterior of a platinum semi-spherical capsule into powdered dry nitre, and fuse the latter, until it forms a central dependent liquid drop over the flame of the suspected gas. On dissolving off the fused nitre by plunging the capsule into a vessel of pure water, and concentrating if necessary, we may prove in it the presence of sulphates by chloride of barium with perfect ease. (See Percy, "Metallurgy of Iron," p. 736.) The presence of manganese in iron ores thus treated appears to increase the greedy absorption of sulphur compounds from the gas flame. Coke, though it usually does not contain more than 0.5 per cent. of sulphur, may contain as much as 3.0 or more per cent., so its contact as fuel is also to be avoided.

The cost of spirit in Great Britain reduces the operator to the use of charcoal only. It is also to be borne in mind that commercial nitre occasionally contains some traces of sulphates.

To determine the quantity of sulphur existing in an ore in the state of a sulphate, we can operate as follows:—

1. If it be sulphate of calcium or sulphate of iron, boil a known weight of the ore, very finely pulverized, in a large quantity of water for a considerable time, filter and ascertain the amount of sulphuric acid contained in the liquid.

[The sulphate of iron may be in the state of basic subsulphate, insoluble in water; therefore, the treatment by water alone will fail to remove the sulphuric acid.]

2. If the sulphate be of barium, and if the proportion be small, we must first treat the ore with acids, filter the liquid, having previously diluted it with water, to avoid dissolving any of the sulphate of barium, calcine the insoluble residuum; then ascertain the amount of sulphuric acid contained in it after its fusion with the alkaline carbonates. In this case we must heat the fused mass with pure water, filter and determine the sulphuric acid contained in the liquid, after having got rid of the silica by evaporation with hydrochloric acid to complete dryness. If the proportion of sulphate of barium be considerable we may fuse the ore directly with the alkalies.

It should be remarked that precipitated sulphate of barium when in presence of alkaline salts (and probably also those of calcium), takes up as much even as 1.5 per cent. or more of these in such a form that they cannot be removed by washing. Hence in the somewhat delicate operation of determining the usually very small amount of sulphur in irons, a grave error in excess may easily occur.

Stolba has pointed out a mode of freeing from these the barytic sulphate (after its washings come off without further action) by digesting it at a boiling heat in a solution of neutral acetate of copper, with some acetic acid added, and then washing afresh, etc.

Volumetric methods for determinations of sulphuric acid have also been pointed out, but they are probably deficient in delicacy for use in the present case.

IX.—ASSAY FOR OXIDE OF ZINC.

First Method: by Hydrosulphuric Acid in an Acetic Solution.

The best method is to precipitate the zinc as a sulphuret by means of hydrosulphuric acid in a solution of the ore containing no other acid than acetic acid in slight excess. The ore being not directly attackable by this acid, we must produce the requisite solution in an indirect manner. Dissolve the ore in concentrated hydrochloric acid, with the addition, if necessary, of nitric acid; add to the solution enough of sulphuric acid to saturate all the oxides it contains, then evaporate to dryness. In this manner we obtain a residuum formed exclusively of the insoluble matters contained in the ore and of the sulphates. Heat with water slightly acidified by sulphuric acid, and pour into the hot solution some acetate of barium until a precipitate ceases to form, allow the precipitate to settle, wash by decantation several times in boiling water, and filter. The clear liquid thus obtained contains all the oxides of the ore in solution in acetic acid and a small quantity of free acid. Pass a current of hydrosulphuric acid through the liquid until it be saturated; the zinc only is precipitated;* it is in the state of a sulphide, and must be allowed to deposit, the air being excluded. Wash by decantation, and then throw the precipitate on a filter, washing it with water containing some hydrosulphuric acid in solution; we must *not* employ the hydrosulphide of ammonium, which would precipitate the iron and other metals present. When the precipitate is sufficiently washed, redissolve it on the filter itself in the least possible quantity of hydrochloric acid, taking care to cover the glass funnel with a watch-glass or plate, to prevent the chance of spurting over by the gas disengaged; wash well and heat the solution to drive off the hydrosulphuric acid; filter and precipitate the zinc again by adding gradually to the boiling liquid a solution of carbonate of sodium, avoiding too great an excess; as soon as the flocculent precipitate becomes granular cease heating, and filter when the precipitate is well collected.

The carbonate of zinc thus obtained is washed, dried, and calcined, carefully avoiding all reducing agents which, owing to the volatility of zinc, might involve perceptible loss. By calcination the carbonate of zinc is changed into the oxide, and this is then weighed. This method is even applicable to ores containing manganese.

---

* If the ore contain copper, lead, or arsenic, besides zinc, we should first get rid of these elements by passing a current of hydrosulphuric acid through the hydrochloric solution of the ore.

Second Method: by the Hydrosulphide of Ammonium in an Ammoniacal Solution.

If we have to treat ores entirely free from manganese, we may use a much simpler process: the solution of the ore in hydrochloric acid containing the iron as a peroxide compound, is exactly neutralized by carbonate of ammonium (a slight precipitate is no inconvenience), then boiled and mixed with acetate of sodium or succinate of ammonium. When the precipitate has collected, filter and wash well with hot water. The liquid being received in a flask capable of being perfectly closed, pour in a slight excess of hydrosulphide of ammonium, and let it rest, access of the air being excluded, as the carbonic acid contained therein might cause precipitation of the lime as carbonate. When the precipitate is collected, decant the supernatant liquid, and then throw the precipitate on a filter, and terminate the analysis as before described.

We may also determine the zinc either volumetrically (see the authors' Supplementary Notes), or in ores containing manganese, by first getting rid of the peroxide of iron and alumina by acetate of sodium, then the manganese by some one of the methods already described, and then precipitating the zinc, as above, by the hydrosulphide of ammonium.

X.—ASSAY FOR LEAD.

First Method: by Hydrosulphuric Acid.

Treat the finely pulverized ore with concentrated hydrochloric acid, evaporate the solution to dryness, and, having moistened the residuum with a little of that acid, pour upon it a boiling solution of chloride of ammonium, in which chloride of lead is soluble. Filter to separate the insoluble matters, wash the precipitate first with warm water, then with a solution of tartrate of ammonium, in order to dissolve the sulphate of lead, if any has been formed; lastly, pass a prolonged current of hydrosulphuric acid through the diluted liquid. A precipitate of sulphide of lead with free sulphur is formed, the latter arising from the reduction of the perchloride of iron. Boil and filter. After desiccation, the filter is introduced into a small weighed porcelain capsule, and a few drops of fuming nitric acid added; then two or three drops of sulphuric acid and a moderate heat applied, until the liberation of vapor has ceased; then increase the heat slowly, until the capsule is red hot, in order completely to destroy the material of the filter. The sulphide is thus changed into a sulphate, the weight of which we ascertain at the balance.

Second Method: by Sulphuric Acid.

Attack the ore with aqua regia, then evaporate the solution to dryness, after the addition of some drops of sulphuric acid. Pour a boiling solution of tartrate of ammonium (which dissolves sulphate of lead) upon the residue; filter to separate the insoluble matters, washing first with a boiling solution of the above tartrate, and then with hot water, and then reproduce the sulphate of lead by decomposing the tartrate of ammonium, which held it in solution by dilute sulphuric acid; add a little alcohol to the liquid to prevent any sulphate of lead remaining in solution, let the precipitate deposit, wash it by decantation, with cold water slightly acidified with sulphuric acid, and mixed with alcohol, and, finally, transfer it by the help of the least possible quantity of liquid into a weighed little flask or glass capsule; decant the greater part of the liquid carefully, and evaporate the remainder to dryness in the weighed vessel, and so determine the weight of the sulphate of lead. If the operation be well performed, the precipitate should be almost perfectly white.

XI.—ASSAY FOR ARSENIC ACID.

Dissolve a sufficient quantity of ore in aqua regia, then reduce the arsenic acid and the perchloride of iron to a lower state of oxidation by means of sulphite of sodium in presence of an excess of hydrochloric acid, keeping the liquid at boiling temperature. When the resulting sulphurous anhydride is completely expelled, dilute with water, let the liquid become quite cold, and then pass into it a slow current of hydrosulphuric acid.

By this operation the arsenic acid was changed into arsenious acid, and the perchloride into the protochloride of iron. In this way the precipitation of the arsenic is facilitated, and a large deposit of free sulphur, which results from a persalt of iron, is prevented. Filter, and treat the precipitate, which is composed of silica, sulphide of arsenic (As^2S^3), and sulphide of lead, if the ore has contained lead, still on the filter, with ammonia, which dissolves only the sulphide of arsenic. Evaporate to dryness the ammoniacal solution in a porcelain capsule over the water-bath, and treat the residue with fuming nitric acid, taking care to keep the capsule covered with a watch-glass or glass plate. As soon as the action of the acid slackens, heat again over the water-bath. The sulphide of arsenic is broken up by this process into sulphuric and arsenic acids.

Dissolve at the same time in nitric acid an accurately weighed quantity of iron, equal at least to half the weight of the arsenic acid presumed to be in the solution. When the iron and the sulphide are completely dissolved, dilute with water and precipitate with ammonia. The precipitate contains all the arsenic acid combined with the peroxide of iron. Filter, wash the precipitate, dry it well, then calcine, taking care to heat it gradually at first, in order to get rid of the ammonia which it may contain, and which, without this precaution, might reduce some of the arsenic acid. We know the amount of the peroxide from the weight of the iron employed, and we thus obtain the weight of the arsenic acid by the difference.

14

[XII.—ASSAY FOR OTHER SUBSTANCES.

Other substances, such as copper, chromium, titanium, cobalt, and nickel, are found in iron ores quite as often as zinc or lead, for which the authors have given methods of determination. We believe, therefore, that it will not be outside of the limits of this work to examine the above named metals briefly.

Copper may exist in iron ores, either combined with sulphur, or arsenic, or in an oxidized state   Its determination is had by passing a stream of sulphuretted hydrogen through the solution of the ore in hydrochloric acid. The solution must not contain too great an excess of acid, and free nitric acid should be especially avoided. The precipitate of sulphide of copper is black, but its color may be changed, if it be mixed with sulphide of arsenic, or the sulphur produced during the reduction of the persalt of iron by the sulphuretted hydrogen. The distinctive colorations will be better perceived if the solution of persalt of iron has been reduced by sulphurous anhydride (or sulphite of sodium) and boiled until no sulphurous smell exists, before passing the stream of sulphuretted hydrogen.

The precipitate is collected upon a filter, and carefully washed with water holding a certain proportion of sulphuretted hydrogen. The washing must be continued without interruption until completed, otherwise a portion of the sulphide of copper may become oxidized and pass through the filter with the liquors deprived of sulphuretted hydrogen. When all the iron has been eliminated, and if it is suspected that sulphide of arsenic has been precipitated with the sulphide of copper, the precipitate is treated with sulphide of sodium (or of potassium) which dissolves the arsenic compound. The hydrosulphide of ammonium is not available in this case, because it may also dissolve a small proportion of sulphide of copper.

The washed sulphide of copper is rapidly dried in the filter, then separated from it as completely as possible, and the filter is burned. The ashes and sulphide are treated with nitric acid to which a small proportion of hydrochloric acid is added, and a moderate heat applied. The copper is dissolved, and the sulphur is wholly or in part transformed into sulphuric acid. It is not necessary that the transformation of the sulphur should be complete; it is sufficient that it separates with a pure yellow color. The acid solution is then diluted with water, filtered, and boiled in a porcelain dish, previous to the precipitation of the oxide of copper, CuO, by a slight excess of caustic soda or potassa. The solution should be rather dilute and boiling before the alkali is added, in order immediately to obtain a dense precipitate of a dark brown color.

The precipitate of oxide of copper is washed two or three times by decantation; and then upon the filter with boiling water, and lastly strongly calcined in a platinum crucible. It is preferable to burn the filter separately, and to add the ashes to the oxide. As the oxide of copper is strongly hygrometric, it should be rapidly weighed after cooling under the drying jar.

Instead of determining the copper as oxide, it is frequently weighed in the form of cuprous sulphide Cu^2S. In this case, the precipitated sulphide is, when dry, mixed with an excess of sulphur, and calcined in a porcelain crucible or tube, in an atmosphere of hydrogen.

Should the proportion of copper be too small to be ascertained quantitatively, the filter and its contents are calcined in a porcelain crucible, and the residue is treated with a few drops of nitric acid. An excess of ammonia will then produce the characteristic blue color due to the presence of copper.

Chromium is often present in magnetic iron ores; and it is generally considered that it is there as sesquioxide of chromium, combined with ferrous oxide.

A qualitative test can be made by calcining, at a red heat, the very finely powdered ore with a mixture composed of equal parts of nitrate of potassium and carbonate of sodium. The oxide of chromium is transformed into chromate of sodium or potassium, which is separated by boiling water and filtration from the iron and other substances insoluble in alkaline solutions. After supersaturation of the alkaline liquor by means of acetic acid, some acetate of lead is added, which produces the characteristic precipitate of chromate of lead (chrome yellow).

If the ore contains manganese, the alkaline solution may be more or less colored in purple by the permanganate produced during the calcination. The manganese will become precipitated by boiling the liquor with alcohol.

The best quantitative test for chrome iron is as follows: mix the highly comminuted ore with about twelve times its weight of acid sulphate of potassium, in a large platinum crucible, which is heated carefully. At first, the temperature should be just sufficient to insure the tranquil fusion of the acid sulphate of potassium, and the equal division of the ore through the molten mass. When this point is reached, the temperature is raised to a dark red heat, which causes the attack of the chrome iron, and the regular and slow escape of the excess of sulphuric acid. This period should last about half an hour, that is, until a little after the whole mass has become thoroughly homogeneous and in a state of tranquil fusion, and no gritty undecomposed substance felt in the crucible by means of a platinum wire or spatula. The temperature is then increased in order to expel all the excess of sulphuric acid, and even partly decompose the sulphates of chromium and of iron.

This first treatment has resulted in the complete disintegration of the chrome iron, which is then in a state to be readily oxidized by a mixture of equal parts of nitrate of potassium and carbonate of sodium. About twelve parts of this mixture to one of ore are employed and put into the crucible holding the sulphates. It is recommended to add at once the whole of the carbonate of sodium with a portion of the nitrate of potassium, the remainder of the nitrate being gradually added at intervals from ten to fifteen minutes. A dull red heat is maintained constant during the oxidation until the whole of the nitrate has been added and the action has slackened. The temperature is then raised and kept up until the mass is in a state of rather tranquil fusion. The whole operation lasts about two hours.

The crucible should be large, so as not to be filled to more than one-half or two-thirds of its height, and, according to the supposed composition of the ore, no more than one-half to one gramme is employed.

The chromates are dissolved with boiling water, filtered, and evaporated nearly to dryness with an excess of nitrate of ammonium, and until all smell of ammonia has disappeared. The alkalies are neutralized by the nitric acid of the nitrate of ammonium, and small quantities of alumina, silica, manganese oxide, etc., become precipitated, and are separated by filtration.

We now have a neutral solution of chromate of sodium (or potassium, or both), the chromium of which may be determined as chromate of lead, or as sesquioxide of chromium. In the first case, the neutral solution is rendered acid by means of acetic acid, and the chromium is precipitated by acetate of lead. Nitrate of lead may also be employed in dilute solutions. The yellow precipitate of chromate of lead is collected and weighed upon a filter previously dried and weighed in the manner indicated at the beginning of this work.

In the second case, the neutral solution of alkaline chromates is rendered strongly acid by hydrochloric acid, and the chromic acid is reduced to the state of sesquioxide of chromium by ebullition with alcohol. This process is slow with diluted solutions, for which it is preferable to employ the agency of sulphuretted hydrogen or of sulphurous anhydride (or sulphite of sodium). The sulphur resulting from the sulphuretted hydrogen is separated by filtration, after thorough settling.

In either case the solution now contains a sesquichloride of chromium which is precipitated as sesquioxide of chromium by a slight excess of ammonia. This precipitate is gelatinous and difficult to wash, like alumina produced under similar circumstances. During its calcination a phenomenon of incandescence takes place, which requires that the crucible should remain covered, until the contents have become thoroughly red.

As a check to the operation, it is necessary to dissolve in hydrochloric acid the insoluble residue, remaining from the filtration of the contents of the crucible in which the attack and oxidization took place. Should there be any dark portions refusing to be dissolved, they are chrome compounds which require a further treatment by the acid sulphate of potassium, etc. This search is to be made as soon as the alkaline chromates have been filtered off, in order not to have to duplicate the further operations.

Titanium in iron ores forms combinations regarding the composition of which there are differences of opinion. Some persons maintain that sesquioxide of titanium is combined with ferric oxide; others believe that the union is between titanic acid and ferrous oxide. It is, however, customary to determine titanium in the form of titanic acid.

The natural compounds of titanium are, some soluble in hydrochloric acid, others not ; but the latter may be rendered

14*

soluble in hot and concentrated sulphuric acid, or by fusion in the acid sulphates of potassium, sodium, or ammonium.

The acid solutions of titanium, when very dilute and boiled for a certain length of time, permit the titanic acid to precipitate in the shape of a powder, which is white and quite pure, if the iron present is in the ferrous state, but yellow or red and very impure if the iron is in the ferric state. The precipitate from hydrochloric solutions is difficult to wash, passes through the filter or obstructs it, and it is not possible to precipitate the whole of the titanic acid by ebullition alone. On the other hand, the precipitation from sulphuric solutions is complete, and the precipitate is easily washed. Sulphuric acid, or the acid sulphates, are therefore to be preferred as means of solution, when the precipitation of titanic acid is to be effected by ebullition.

If, for some reason or other, it is not thought desirable to employ sulphuric acid or its compounds, the titanic substance (when undecomposed by hydrochloric acid) is fused with carbonate of sodium, or the sodic carbonate of potassium, and then dissolved, *in the cold*, in hydrochloric acid. In the case of such a solution holding only iron and titanium, the two metals may be precipitated together, in the cold, by ammonia ; the oxide of iron is transformed into sulphide by hydrosulphide of ammonium, and the sulphide is rendered soluble by an addition of sulphurous acid. The titanic acid which remains is gelatinous, and resembles alumina precipitated under similar circumstances. After washing, it is dried, calcined, and weighed.

The attack by sulphuric acid requires that the ore should be very finely powdered. The acid used is a mixture of about equal volumes of sulphuric acid at 66° Bé. and water. The heating is begun on a water-bath, and continued at an increased temperature until the ore is entirely decomposed, and there are produced abundant fumes of sulphuric acid. The fluid mass, which is often syrupy in appearance, is allowed to cool, and is then diluted in a great volume of water, avoiding an elevation of temperature, which may cause the precipitation of a part of the titanic acid. The dilute, clear, and cold solution is filtered, if silica be present, and the iron is transformed into a protosalt by sulphurous anhydride, or by an alkaline sulphite. By boiling the liquor for a long while, and keeping it smelling of sulphurous anhydride, and making up the loss of evaporation by hot water, all the titanic acid is precipitated as a white powder, easily washed. The more free acid there is in the solution, the more it should be diluted. If the precipitate, dried at 100° C., continues perfectly white, it may be considered as free from iron. If yellow or brown, it should be dissolved anew in strong sulphuric acid, and treated as before. The calcined titanic acid seldom remains white, although pure ; it becomes yellow, gray, or brown according to the intensity of the temperature applied.

The solution filtered from the titanic acid contains the other substances from the ore, which are determined in the ordinary manner. The silica, separated before boiling, retains sulphuric acid with such tenacity, that it is recommended to fuse it with carbonate of sodium, and then to follow the usual treatment with hydrochloric acid, etc.

The titanic acid retains also sulphuric acid with great tenacity ; therefore, after the first calcination, a small lump of carbonate of ammonium is put in the cold crucible to remove the acid, and the temperature is slowly raised to a red heat.

The attack of the titaniferous ore by the acid sulphate of potassium is, in the writer's opinion, more easy than that with sulphuric acid. The operation is conducted in a manner similar to that explained for the disintegration of the chrome-iron ores, but the heat should not be raised so high as to expel all the excess of sulphuric acid and decompose the sulphate of iron which must remain soluble. From ten to fifteen parts of acid sulphate of potassium are employed to one of ore, and the whole of the sulphate may not be employed at once. When the swelling of the material in the crucible has subsided, and the fusion is tranquil, the temperature is raised to a low red heat, and kept so until the whole of the ore has become dissolved, and no grit is felt with a platinum spatula or wire. The temperature may then be increased a little more for a few moments. After cooling, a fresh quantity of acid sulphate of potassium is added, and the heat is again applied, and gradually raised until the whole mass is homogeneous and still emits fumes of sulphuric acid. The crucible is then allowed to cool. In this manner, all the free acid is not removed, and we are certain that the iron, alumina, etc., are in the state of sulphates, and soluble.

The cold crucible and contents are digested in cold water. The solution requires several hours, and may be hastened by stirring. The remainder of the operation is the same as when sulphuric acid is employed.

Cobalt and nickel require in this place but a notice as to their qualitative determination. They are frequently together, and in small proportions in certain kinds of iron ores, and are not considered injurious to the metal. They are not precipitated by hydrosulphuric acid, carbonate of barium, or by the usual treatment with acetate of sodium, although a small proportion may remain with the other precipitated metals. On the other hand, they form with the hydrosulphide of ammonium black sulphides, which are not soluble in dilute hydrochloric acid.

Therefore, if in a solution of an ore deprived of copper, lead, or arsenic by a previous treatment with hydrosulphuric acid, and of iron and alumina by the carbonate of barium or the acetate of sodium, we add hydrosulphide of ammonium, we will obtain a black precipitate of the sulphides of cobalt or nickel (should these substances be present), which may be mixed with the sulphides of manganese and zinc. The latter are rendered soluble and separated by adding hydrochloric acid, drop by drop, and the sulphides of nickel and cobalt remain.

If a special test is made for these metals, it is not necessary to submit the solution of the ore to the treatment by carbonate of barium or acetate of sodium. The metals unacted upon by hydrosulphuric acid are precipitated together by the hydrosulphide of ammonium, and the sulphides of nickel and cobalt are separated by dilute hydrochloric acid.

An approximative quantitative determination may be had, by dissolving these sulphides in nitric or nitromuriatic acid, and precipitating the nickel and cobalt together as oxides, by means of caustic potassa.]

ANALYSIS OF SCORIÆ, SLAGS, FLUXES, AND FIRE CLAYS, ETC.

The elements of these substances are much the same as those which enter into the composition of ordinary iron ores. The iron exists in scoriæ and slags chiefly in the state of protoxide; the remainder, as peroxide, free or combined—sometimes even metallic iron is found in disseminated particles. The analysis of these matters does not usually require any special methods; attention should chiefly be given to the evaporation necessary to render the whole of the silica insoluble, and to the presence of the peroxide of iron.

A little should be added to the text here. Blast furnace slags are very generally, though not always, pretty easily brought to a soluble state by fusion with the alkaline carbonates. Some, however, as well as some scoriæ (from the wrought iron or steel manufacture), are very refractory. In some scoriæ, as in those from the puddling furnace, the percentage of phosphorus is high, and may be an object of industrial research hereafter, though not as yet so. Hence it is desirable to employ no method with these likely to lose part of that element. Again, in the analysis of fire clay, or fire brick, it is of great importance to determine the amount of alkalies naturally present, which seriously affect their infusibility. For these last bodies, fusion with the alkaline carbonates is inadmissible.

Professor Lawrence Smith's method by fusion with a mixture of pure fluor-spar and carbonate of calcium is stated to give good results, and is easily practised (see Crooke's "Methods of Chemical Analysis"), and like the methods of Brunner and Laurent with hydrofluoric acid, or the modification by Gore, which consists in melting with nitrate of barium and fluoride of barium, may be employed for scoriæ slags or fire clays, and these all admit of the determination of the contained alkalies.

[The attention of the beginner is called to the fact that slags or cinders contain the sulphur, phosphorus, and arsenic, in the state of sulphates and sulphides, phosphates and phosphides, arsenates, arsenites, and arsenides, and that the mode of treatment is to vary as it is desired to determine these substances in their various forms of combination. For practical work, however, it is sufficient to determine these bodies as sulphur, phosphorus, and arsenic.

Certain kinds of blast furnace cinders contain small copper-colored crystals of nitrocyanide of titanium.

It would also be desirable to ascertain whether the oxide of aluminium contained in blast furnace cinders, is, or not in a lower state of oxidation than alumina, Al^2O^3.]

PART V.

---◆---

ASSAY OF IRON ORES BY THE DRY METHOD.

THE assay by the dry method serves, in the first place, to make known the yield in cast-iron of an ore; in the second, it controls the management of the blast furnace by fusion in the crucible of a mixture identical with that employed on a great scale; thirdly, it enables us to compare to a certain point the quality of the cast-irons produced by different ores; lastly, it may have no other object than to procure a button of cast-iron, intended for the assay of its contents in phosphorus, as we observed when speaking of that particular body.

The basis upon which this assay rests is as follows: If we put a certain quantity of iron ore into a clay and charcoal-lined (*brasqué*) crucible, with fluxes, and expose the mixture to a progressively increasing temperature, the volatile products, water and carbonic acid, are first liberated; at about 400° Cent. the proto- and peroxides of iron in immediate contact with the clay and charcoal-lined crucible are reduced, with production of carbonic oxide and carbonic anhydride, which the excess of carbon present reduces to carbonic oxide. The carbonic oxide, diffused through the mass, and changing into carbonic anhydride, reduces in its turn the part of the ore not subjected to the direct action of the *charcoal* lining of the crucible. The reduced metallic iron is carburetted either by the gases or by the solid carbon. At a temperature of about 1500° Cent., the cast-iron collects gradually in the bottom of the crucible, whilst the earthy matters, becoming vitreous, melt in their turn, and form the scoria that rises to the surface above the cast-iron, which, when cooled, separates as a metallic button. At the temperature at which the cast-iron is produced, zinc and lead are volatilized, and are not found in the button or ingot; but this is not the case with the manganese, sulphur, and phosphorus. If the ore contain these elements, they combine at least in part with the iron as well as does also a certain quantity of silicon.

Mr. J. Lowthian Bell, and M. Gruner (of the *École des Mines*, Paris), have both shown that at temperatures of 300° to 400° Cent. carbonic acid in presence of iron ores is split up into carbonic oxide and carbon, provided carbonic oxide be also primarily present.

It follows, therefore, that the reactions during the reduction of iron ores are not always and quite those described in the text. Mr. Bell's work on "The Chemical Phenomena of Iron Smelting," and M. Gruner's Memoir ("Comptes Rendus," 73, 88, 1871) should be consulted.

The assay is generally made in crucibles from 10 to 12 centimetres high, of good refractory clay, and charcoal-lined (*brasqué*).

No better crucibles for iron assay are to be met with than those made for the purpose in Cornwall, chiefly about Redruth. They are broader in form than French or German crucibles, and stand sudden changes of temperature very well, and are highly refractory.

Nor is any form of assay furnace, the writer believes, better than the South Wales and Cornish pattern, which is distinguished by the great thickness given to the front wall so as to protect the operator from the heat, which in the case of thin-walled furnaces renders the continuous work of the assayer unpleasant and unhealthful.

Plumbago crucibles are sometimes employed, and are suggested in the larger books on Metallurgy; but they are expensive, and the recommendation that one may be used more than once is of more than doubtful value. Iron assays are frequently made in Great Britain in unlined crucibles, as much carbon in the state of pulverized coke or anthracite being mixed with the ore and flux as may be required for the reduction. The writer, however, deems the charcoal lining of the crucible better practice, and its result is generally to produce a better *button* and fewer detached particles of cast-iron in the assay. The method of *brasquing*, or charcoal lining, as described in the text, gives good results when well performed; but the crucible should be dipped into water before the charcoal *dough* is pressed in, otherwise it is but moderately adherent. A very perfect form of charcoal lining consists in making the charcoal, with water and a very little molasses, into a thickish cream, pouring some into the damped crucible, inverting the crucible until no more will run out, and then drying slowly in an oven, and repeating this until a sufficient thickness of lining is obtained. Where lined crucibles are adopted, a stack of dried ones should be maintained ready for use.

For the lining wood charcoal only is used. It is finely powdered, passed through a wire-gauze sieve, and wetted with a little water carefully mixed through it by kneading. The quantity of water should be barely that required to knead the mass under considerable pressure. A small quantity of the mixture is put into the bottom of the crucible, then pressed down with a pestle or piece of wood

15

specially rounded at the extremity; to this first layer a second is added, then a third and fourth, until the crucible is filled. To prevent the different layers being independent of each other, we should scratch the surface of each layer with a blade or point of iron before putting in another. The solid centre is then scooped out with a knife; the cavity should only descend about two-thirds of the interior depth of the crucible, and a regular thickness, equal to about the fourth of the diameter, should be left all round. The surface of the cavity, and the bottom especially, should be very smooth. A thick glass tube, closed and rounded at the extremity, is used to effect this smoothing or even polishing of surface.

The advantage of the test by the dry method is to furnish results analogous to those obtained in the blast furnace. To·render this analogy as complete as possible, we should exclusively employ as fluxes the substances used in operating on the large scale, and systematically reject any others, such as soda, borax, etc. Lime, alumina, and silica are the only fluxes made use of.

Lime is employed in the state of carbonate; instead of chemically pure carbonate, good white limestone is generally used, free from pyrites, phosphates, and metallic oxides. It is pulverized finely in tolerably large quantity, and its composition being ascertained, it is reserved for use. A small proportion of magnesia in the limestone is not injurious.

The alumina is employed either alone, as found in the pipeclay of commerce, or hydrated, and in combination with silica, in the state of potters' clay. In the latter case a very aluminous clay is selected, dried, finely powdered, and its composition, like that of the limestone, precisely ascertained.

The silica is either in combination with the alumina or separate, but in the latter case it should be in impalpable powder. It is obtained in this form by levigating silicious sand, or by decomposing a solution of an alkaline silicate with hydrochloric acid, evaporating the liquid to dryness, and washing the residue with hot water until nothing but silica remains.

Some assayers employ earthy glasses of known composition, and prepared by fusing in charcoal-lined crucibles, mixtures of potters' clay and limestone.

The general formulæ given in the text for fixing the relative proportions of silica, alumina, and lime in the flux, though correct in principle, may probably be accepted by the practical British assayer as mere learned trifling. They have their *occasional* use, however, when dealing with some new and little-known ore ; but to the experienced assayer there is no doubt that they are generally useless.

Assaying by the dry method, like almost every *art* to which science is the directrix only, soon becomes, under the teaching of experience, an almost mechanical craft, and without this could scarcely become rapidly conducted enough to be profitable when practised as a profession.

Accordingly, nothing can well seem more rough and ready, or almost hap-hazard to the mere looker-on than the operations of the iron assayer, or even of the assayer of copper ores, who deals with more complicated problems.

The several flux materials are kept beside each other in small ranges of wooden boxes, seldom even labelled. The ores, after fine pulverization, are weighed out (usually in Great Britain by assayers' weights, which bear such a proportion to the ton of 2240 lbs. as enables results to be calculated in very few figures ;

but where the French weights are employed the gramme is as convenient as any other unit) and placed each on a numbered paper (or dish, sometimes of glass, more usually of thin copper), and placed in a numbered square upon a tray, with which they are transferred in batches from the balance-room to the table of the assayer's furnace-room, on which the fluxes stand.

The assayer passes the ore from one of the papers or dishes (replacing the latter on the tray) into a copper shovel or shoot, much like those used by bankers for gold coin, but longer and narrower mouthed, and then adds the several ingredients to form the flux, by taking up as much of each as in his judgment is proper, by means of a small bent hemispherical spoon or ladle. If unlined crucibles be employed he adds powdered anthracite in like manner; mixes all tolerably evenly together upon the shovel, and shoots the whole thence into the cold crucibles, which also stand upon a tray divided and numbered to correspond with that for the ores.

The wind or draught furnace is already very hot. He covers up the assay with more anthracite dust, and lutes on the cover, but not infrequently operating without any cover to the crucible except a deep stratum of not too finely powdered anthracite. Usually, however, covers are employed, and these being fitted he puts the crucible, or more than one, directly into the furnace, without any stand or support but what it receives from the good hard cokes, of which he throws in a few fresh pieces just before he inserts the crucibles, so as to damp down the heat for the moment. Good clay crucibles of Cornish make seldom crack even with this rough treatment. The march of the furnace is watched from time to time, so as to see that by the descent of the fuel, as the combustion goes on, the crucibles are not endangered by oversetting. Their position is rectified, if need be, by light, though long-handled, spring tongs, used by one hand only, and fresh fragments of coke are thrown in when needed.

The assayer knows by experience almost precisely the time required for the complete reduction and fusion; and after slightly stirring round with a stick of dry wood, or without any previous operation, he withdraws the crucibles, sways each round a little in the grasp of the tongs, and knocks it against the brick floor, and leaves it to cool, when each is replaced on the numbered tray whence the crucibles were taken, the assayer always preserving (if more than one crucible be heated at once either in the same or in adjacent furnaces) the same order as to position therein, as the cold crucibles occupied upon the tray—thus confusion of assay is avoided. When cold the crucibles are broken separately over a tray, and the buttons and fragments, whether of metal or of slag, placed in their own compartments on the numbered ore tray, for examination, weighing, and record.

We believe there are assayers who still adhere to the crucible stand, and to the placing that and the crucible, or more than one, in a cold furnace, kindled as described in the text; but the rate of progress is slow, and probably there are not more failures by one method than by the other. The mixture and addition by judgment only of the fluxes to the ores of course involves this, that the assayer is working upon classes of ores which he has been accustomed to; and no assayer, however experienced, can be certain of a good assay at the first trial upon an ore very diverse from any he has before operated on. He obtains gradually, however, a tactile sort of knowledge of the properties of the ores habitual in his district, or in commerce, which is frequently marvellous in its accuracy, and led by that is seldom astray in the flux mixture he employs, or for more than once at most.

A good assayer is never satisfied with the result of his assay, unless the whole of the metal be collected into one beautifully smooth and rounded spheroid, taking at its lower side the form of the crucible, or of its lining, and showing at its upper surface that form which capillarity confers while fluid upon a perfectly fused metal; and if the scattered fragments amount to more than a very few *spherical* shots varying in diameter, he rejects the assay, for if the uncollected fragments be so numerous as to require collection with the magnet, it is impossible to be sure we have collected all; and we may collect with minute fragments of iron those of adherent slag.

The assay furnace for iron needs a better draught (if time is to be economized) than that for copper; and perhaps the power of turning on a gentle fan blast from the ash-pit (as done in some bronze founderies, notably that of Messrs. Barbedienne, of Paris) would still further save time. The crucibles should, however, be heated up at first by the natural draught of the furnace alone, as the fan blast would be too brusque for safety.

Very little reliance can be placed upon what judgment may be formed from the fracture, etc., of an assay button as to the physical or mechanical properties of the cast-iron itself. The external qualities of cast-iron are far too readily affected by minute differences of constitution, by more or less rapid cooling, etc., to admit of any such judgment being made with safety; and when a new ore is under consideration for use in the blast furnace, and there be any reason to doubt the quality of cast-iron it may render, the only safe plan is to reduce some 10 to 20 lbs. of the iron in a large plumbago or steel melter's pot, and to pour out the fused metal into a "dry sand" ingot mould, made always of the same dimensions, which may be about 15 inches long by 1½ or 2 inches square, and submit this to examination.

Given an ore, what are the proportions and the nature of the fluxes to be employed for a successful assay? Such is the question we are now to answer.

With respect to the ordinary dimensions of crucibles, we may consider 50 grammes as the maximum weight of the mixture to be fused, and 10 grammes the minimum; for, with a less quantity, the button is too small, and it is difficult to decide as to the nature of the cast-iron; besides, the chances of error increase as the quantity operated on is greater.

The ratio in the weight of the cast-iron to that of the slag produced in an experiment must not be arbitrary; in fact, as the slag always contains some iron, it is obvious that it is injurious to increase its weight unnecessarily; nor should we diminish it below a certain limit, for without slag, or with too small a proportion, the collecting of the cast-iron into a single smooth round button would be difficult or impossible.

We may consider the ratio of one to one as the most expedient to attain, and which should not be exceeded unnecessarily; moreover, there is some advantage, as in operating with this ratio we are in conditions more nearly identical with those employed in the blast furnace. In our choice of the *quality of the fluxes*, we are now guided by the following considerations: The slag should not become fused until after the complete reduction of the oxides of iron; for if it were previously fused, it would carry with it a perceptible amount of these oxides; on the other hand, it must be completely fused at a temperature attainable in the assay wind furnace, for upon the proper fusion of the slag the success of the assay greatly depends. The presence of sulphur in the ore also exerts an influence on the selection of the flux. A sulphurous ore requires proportionally more lime than an ore free from sulphur.

It is difficult to fix the limits of composition within which the slag is still sufficiently fusible. Experience has shown that within the very extended limits admissible, the best results are obtained with slags containing only the silicates of calcium and aluminium, in the following proportions:—

        5 to 15 per cent. alumina.
        30 to 50   "   lime.
        35 to 55   "   silica.

Approaching the higher limit indicated for silica, we obtain a slag which fuses easily; if by a mixture of fluxes we approach the inferior limit, we obtain a more basic slag requiring a much higher temperature for fusion, and consequently having a tendency to produce a highly carburretted cast-iron (gray or mottled iron), and exercising by its large proportion of lime, a desulphurising action on the ore.

Having ascertained by the wet method the exact composition of the ore, we can then calculate the quantity of each of the fluxing materials to be employed so as to obtain a slag of definite composition.

Let a, c, and s, represent the respective proportions of alumina, lime, and silica, contained in 100 parts of ore dried at 100° Cent., and x, y, and z, the minimum quantity of each of the three ingredients which should be added to 100 parts of the ore, in order to obtain a slag containing A, C, and S, per cent. of these same ingredients, we shall have—

$$\frac{a+x}{A} = \frac{c+y}{C} = \frac{s+z}{S}$$

In order that the sum of $x+y+z$ be a minimum, one of those unknown quantities must be $= 0$.

Supposing that this be x, we have—

$$x = 0; \; y = \frac{aC}{A} - c; \; z = \frac{aS}{A} - s;$$

when y is $= 0$ we get—

$$x = \frac{cA}{C} - a; \; y = 0; \; z = \frac{cS}{C} - s;$$

finally when $z = 0$, we get—

$$x = \frac{sA}{S} - a; \; y = \frac{sC}{S} - c; \; z = 0.$$

Of these three systems, one alone will not give a negative value, and it is that one which we require.

The composition of the ore, and the slag to be obtained, being given, the general symbols are replaced by their numerical values, and we use that one of the series which leads to a positive value.

For instance, an ore of the following composition being given:—

Alumina   =   4.47
Lime   =   6.75
Silica   = 14.32
Oxide of Iron = 71.13 (iron = 49.79 per cent.)
Water, etc.   =   3.33
                  ————
                 100.00

Supposing we wish to obtain a basic slag of centesimal composition—

Alumina 14
Lime   48
Silica 38
        ————
        100

We replace the letters a, c, and s, and A, C, and S, by their numerical values, and we get in the three series:—

I.

$$x = 0$$

$$y = \frac{4.47 \times 48}{14} - 6.75 = 8.58$$

$$z = \frac{4.47 \times 38}{14} - 14.32 = -2.19$$

II.

$$x = \frac{6.75 \times 14}{48} - 4.47 = -2.80$$

$$y = 0$$

$$z = \frac{6.75 \times 38}{48} - 14.32 = -8.98$$

III.

$$x = \frac{14.32 \times 14}{38} - 4.47 = 0.81$$

$$y = \frac{14.32 \times 48}{38} - 6.75 = 11.34$$

$$z = 0.$$

We here see that the last system is the one the values in which we should employ.

But there is another question: does the ratio in the weight of the slag, required to that of the cast-iron agree with the quantitative fluxes that we have just calculated, and if not, can we make it agree, and how?

This is expressed in a general way by the formula

$$\frac{a + c + s + x + y + z}{F},$$

F being the amount of iron contained in 100 parts of the ore. If this ratio be equal to unity it is all right, and if it be greater than unity we admit it; but if it be less, we can arrange it without changing the composition of the slag to be obtained, by increasing x, y, and z, the amounts proportional to A, C, and S.

In this case let

$$a + c + s + x + y + z = F - k.$$

The quantities $x' + y' + z'$, to be added in order to make the ratio of the weight of the slag to the weight of the cast-iron equal to unity, are given by the formulæ—

$$x' = \frac{Ak}{100}, \quad y' = \frac{Ck}{100} \text{ and } z' = \frac{Sk}{100}$$

k being equal to $x' + y' + z'$ and $100 = A + C + S$, by the conditions of the problem.

The quantities $x + x'$, $y + y'$ and $z + z'$ of alumina, lime, and silica are those which we would employ, if the assay were made with 100 grammes of matter; but as we always operate with less amounts, we must reduce them in the ratio of 100 to the number of grammes of ore we actually employ.

$y + y'$ being the quantity of lime calculated in the form of oxide of calcium, we must calculate the weight of the carbonate of calcium that this quantity of lime corresponds to.

Let us take the example given above; calculation gives us the values—

$$x = 0.81; \quad y = 11.34; \quad z = 0.$$

The quantities of alumina, lime, and silica contained in the ore are—

$$a = 4.47; \quad c = 6.75; \quad s = 14.32.$$

The quantity of iron corresponding to the oxide is—

$$F = 49.79;$$

the ratio $\dfrac{a + c + s + x + y + z}{F}$ is therefore equal to $\dfrac{37.69}{49.79}$ and $k = 12.10$.

The quantities of the different fluxes to be added in order to make the ratio of the slag to the cast-iron equal to unity, will therefore be—

$$x' = \frac{14 \times 12.10}{100} = 1.69, \, y' = \frac{48 \times 12.10}{100} = 5.81$$

$$z' = \frac{38 \times 12.10}{100} = 4.60.$$

We should then employ—

$$0.81 + 1.69 = \quad 2.50 \text{ of alumina}$$
$$11.34 + 5.81 = 17.15 \text{ of lime}$$
$$4.60 \text{ of silica.}$$

If the calculation be correct, the sum of these three quantities, added to that of the numbers which represent the quantities of alumina, lime, and silica contained in the ore, should give a number equal to 49.79; that is to say, equal to the weight of the iron it contains.

In consequence of the very large limits admissible for scoria, we can soon acquire the habit of judging from looking at the ores with which we have most to do, what fluxes should be added. For very clayey ores, an addition of lime only will suffice; with calcareous ores we must employ silica or alumina, or a mixture of the two; according as the ore contains, in addition to lime, a more or less large proportion of alumina and silica; lastly, with ores consisting of almost pure oxide of iron, we should use a mixture of lime, alumina, and silica, or of glass, these prepared beforehand, remembering what has been said with regard to the proportion to be maintained between the cast iron and the slag.

Even if we do not approximately know the composition of the ore, we may obtain a good result by guess; for example, making a series of experiments with the addition of 5, 10, 15, 20 per cent. of lime, if none of these succeed, we should commence a second series with the addition of silica or alumina; the first generally enables us to recognize what constituents are wanting for the slag, and we are almost certain to succeed in the second trial.

But we are seldom reduced to this extremity, for a few simple quantitative trials give sufficient indications of the composition of the ore.   For instance, if the ore effervesce with acids it usually contains lime; if it be very clayey it will emit, when breathed upon, the characteristic "clayey" odor, and from the residue when treated with hydrochloric acid, and often by mere pulverization, we ascertain whether it contains much silica.

When we have selected the fluxes, we mix them well with 10 to 20 grammes of the ore on a sheet of glazed paper, and place them carefully in the bottom of the cavity scooped out of the charcoal lining of the crucible, bringing its particles closer together by knocking the crucible lightly on the table, and filling up the remainder of the cavity with powdered charcoal.   This done, we arrange a cover for the crucible, provided with a small opening for the liberation of vapor and gas; set it on its stand, and unite all the parts with fire-clay, and place the stand on the grate of the furnace.   In case we wish to heat

16

several crucibles simultaneously, we should take care to arrange them so as to allow a sufficient space for fuel between them; otherwise the success of the assay might be compromised in one or more. It is very important to manage the fire well, and to heat the crucibles gradually. Coke is generally employed; it should be in pieces sufficiently small to fill the interstices between the crucibles; if it were too small, it would prevent the free circulation of the air necessary for a powerful heat—or if too large may let pass currents of cold air. The crucibles being placed on the grate, we surround them with unignited coke in quantity sufficient to cover them completely. On the coke, throw some well-kindled charcoal, close the furnace, and let the coke kindle equally; then fill the furnace with fresh coke to the level of the flue leading to the chimney stalk. After some time, the fire having gradually extended downward to the grating, we pack the coke with the help of an iron rod, and, if necessary, refill the furnace with fuel, and keep up the fire well until it may be let to burn out. We can withdraw the crucibles from the furnace by a pair of tongs while still hot, and by shaking each and gently knocking it on the floor, collect the grains of iron which may be scattered through the slag, but it is generally preferred to let the crucibles cool in the furnace. When cold, we endeavor to knock off the cover without breaking the crucible; but if unsuccessful, we break it in two by a blow of a hammer, and detach and collect the contents carefully upon a sheet of glazed paper. If the operation be well performed, we shall find the iron and the slag forming a single well-fused mass, the iron in a button at the bottom, and easily detached clear of slag. The metallic button does not always represent the sum total of the iron obtained; usually, the slag contains some more or less spherical grains, which we must collect; for this purpose pulverize the slag, and withdraw the metallic particles by means of a magnet. We look for the same in the remaining charcoal lining, and when all the particles are collected we weigh the whole of the metal. From this weight we can calculate the percentage of iron furnished by the ore assayed. This proportion should always be in excess of the proportion of iron calculated from the results of an assay by the wet method, for cast-iron contains, in addition to iron, carbon and silicon, and also (if the ore possess these elements), sulphur, phosphorus, and manganese, and many ores contain some other bodies. It is true there is always a small quantity of iron left in combination with the slag, but when the assay has been well performed, that is very small, and the loss due to this cause does not compensate for the increase of weight caused by the presence of the carbon and the other bodies we have mentioned.

If the operation be successful, the exterior surfaces both of the iron button and of the slag should be smooth. Rough surfaces in the iron button, a porous texture in the slag, and the partial dissemination of iron through it, are proofs that the assay has failed, or is unreliable either from insufficient heating, or a misproportioning of the various fluxes employed. Little can be inferred from the appearance of the slag, for assays made with the same ore and the same fluxes in the same proportions, may give slags of widely different aspects, according to the temperature to which they have been subjected, and the more or less rapid cooling of the crucible. We may consider a bottle-green and vitreous slag as containing oxides of iron, which may be attributed to a large proportion of silica in the ore or flux, and hence to the too great fusibility of the mixture; and we may regard apple-green slags, opaque at the surface, or even all through, as indicative of the presence of manganese.

The aspect and texture of the cast-iron varies not only with the nature of the ores employed, but to a certain degree with the manner in which the assay has been performed, with the nature of the fluxes, with the length of time during which the metal has been exposed to the action of heat, and with the temperature to which it has been raised. We should, in the first place, examine the cast-iron button, with respect to its malleability, trying how far it may be hammered out without cracking at the edges or breaking, and then we must judge of the grain and color from the fracture. Malleability is in general the mark of a good quality. Cast-iron is commercially classified, from these and other data, as "gray," "mottled," "white," and "specular" (*spiegeleisen*).

The last variety is only obtained from ores containing much manganese, which furnish the best quality, when a large proportion of lime is employed, the effect of which is to cause more of the reduced manganese to combine with the cast-iron. Manganese appears to assist in purifying cast-iron by sending more or less of the sulphur and phosphorus into the slags.

Ores containing a large amount of phosphoric acid generally give mottled or white cast-iron.

Neither by one nor by the other, nor by any method, is it possible to predict the proportion of phosphorus which shall be contained, on the average, in either pig or wrought iron produced from a given ore, or mixture of ores, by the ordinary methods of manufacture on the great scale. Usually, the ores not being excessively rich in phosphorus, mere traces are found in the blast furnace slags; but when the ores are highly phosphoric, and much unreduced, iron remains in the slag, the latter contains sensible amounts of phosphorus, and the iron less per cent. than is due to the ore.

The fuel, temperature of blast, flux, and burden, as well as dimensions of blast furnace, all influence the proportion of phosphorus in the pig-iron. Berthier's Memoir, in "Annal. des Mines" for 1838, should be consulted.

[A dry assay of iron requires about two hours, one for the period of reduction, and one for the fusion. There is no inconvenience in lengthening the first period, but the fusion should be rapid, and at the highest temperature possible without fusing the crucibles. The color of the cast-iron button varies with the mode in which the firing has been conducted. There is very little analogy between the proportions of carbon, manganese, and silicon obtained in this assay, and those of the same metals in the pig-iron produced in the blast furnace. If the button is of bad quality, it may be interesting to analyze it for sulphur, phosphorus, and arsenic. The dry assay has lost much of its importance, since the wet methods of analysis have been improved.]

16*

PART VI.

———•———

ANALYSIS OF CAST-IRON, MALLE-
ABLE IRON, AND STEEL.

THE silicon, phosphorus, sulphur, and manganese, which these metals may contain, as well as the carbon, which is a constant constituent, have a very great influence on their industrial qualities. It is not then surprising that the value of analysis should be more and more appreciated in reference to their commercial relations. The time will arrive when sales of these metals will not be effected without a certificate of their analysis. The great want of analytical methods furnishing precise results, yet with a rapidity sufficient for industrial purposes, is probably the principal, if not the only, cause restricting such a course.

Some remarks are to be made with regard to the selection of the specimen for analysis. According to the mode of analysis employed, it may consist of a single fragment, or of some shavings, granules or filings. The hammer, the graver, the common mortar, Abich's mortar (otherwise called the steel or diamond mortar), and the file, are used, according to circumstances, to bring the material to a finely-

divided form. The drilling or planing machines
might be advantageously employed for this purpose.
The pulverization of cast-iron containing graphite in
an open mortar may, by loss of uncombined carbon,
occasion a very material error which should certainly
be avoided. This is occasioned by the motion of the
air, which tends to carry off the finest and lightest
scales of the graphite. Pending the operations
which prepare the specimen for analysis, especially
during the sifting, care should be taken to exclude
organic matter, the presence of which would falsify
the determination of the carbon. A metallic sieve,
made of copper wire gauze, or a thin sheet of copper,
pierced with very fine holes, is best for this purpose;
hair or silk sieves are unsuitable.

The complete analysis (in an industrial point of
view) of a specimen of cast-iron* includes the determination of manganese, silicon, phosphorus, sulphur,
carbon (in both states), and iron itself, which last we
must not neglect, if we desire to control the results
obtained by analysis as to the other elements.

I.—DETERMINATION OF IRON.

Dissolve from 0.5 gramme to 1 gramme of the
specimen by means of slightly diluted hydrochloric
acid, to which add from time to time a few crystals

---

* Under the name cast-iron ("*fonte*") we shall include in the
subsequent pages malleable irons and steels, unless when we find
it necessary specifically to refer to either of these.

of chlorate of potassium, and assist the reaction by heat. When the solution is effected, dilute with a little water, and filter to separate the insoluble residue. If this still contain iron, which is almost always the case, slightly wash and filter and burn it in a platinum crucible, then fuse the product of this calcination in the same crucible, with a small quantity of sodic carbonate of potassium, and a little saltpetre. The fused mass is treated with diluted hydrochloric acid, and the solution thus formed is added to the first liquid. Heat to expel the free chlorine, and ascertain the amount of iron by means of chloride of tin (as previously directed). If permanganate of potassium be preferred, it is necessary to add first to the liquid a small quantity of sulphuric acid, then evaporate to dryness, treat with dilute sulphuric acid, and reduce the persalt of iron to the protosalt by means of zinc.

II.—DETERMINATION OF MANGANESE.

We operate upon a specimen larger in proportion, as the manganese supposed or indicated to be in the iron by a previous experiment is smaller. Make the solution as directed in the preceding paragraph. As the filtered liquid contains the whole of the manganese of the specimen, there is no occasion to notice the insoluble residue. Ascertain the quantity of manganese according to one of the methods already detailed in reference to the analysis of ores. Observe, that a single operation is not sufficient to separate manganese completely from the iron.

III.—DETERMINATION OF SILICON.

Silicon is found in cast-iron, partly free and partly in combination, as silicide of iron, partly as silicic acid in combination with bases, in the form of particles of slag intermingled with the cast-iron. To ascertain the quantity of silicon, dissolve 5 to 10 grammes of pulverized or simply granulated cast-iron in slightly diluted hydrochloric acid. A porcelain dish may be employed, but better still a platinum capsule, for porcelain is not always safe from attack by acids. After complete solution, add some drops of sulphuric acid to the liquid, then evaporate to dryness in the water bath until hydrochloric acid is no longer liberated. The silica is rendered insoluble. Treat hot with hydrochloric acid and water, and separate the insoluble residue, which contains different substances, by filtration. In order to isolate the silica, burn the filter, and fuse the residue with the alkaline carbonates and a little saltpetre; treat the fused mass, to obtain solution, with hot water containing a little hydrochloric acid and some drops of sulphuric acid; evaporate to dryness in the waterbath. Again dissolve it with some drops of hydrochloric acid and hot water; everything is now in solution, except the silica, which is separated by filtration, dried and weighed after calcination. This weight corresponds to the total silicon; to ascertain in what proportions this is divided, between free silicon and the silicon involved in the cast-iron in the state of slag, we must determine the latter in the manner about to be described, and the former by the difference.

Determination of Silicon in the Involved Slag.

We are not aware that more than one method for this has been proposed, viz., that given by Frésénius in his "Treatise on Analysis," which we shall give verbatim, with the manner in which that author recommends the solution to be made.

"In order to ascertain the amount of slag involved in cast-iron, dissolve a sufficient portion in very largely diluted hydrochloric acid, with the help of a voltaic current. The solution is effected with the feeble current of electricity, obtained by a single element of Bunsen's, by plunging the piece of iron to be analyzed as the positive electrode into the diluted hydrochloric acid. The iron is dissolved as a protochloride, giving up its carbon, and forming no gaseous product with it and the hydrogen liberated at the opposite or negative electrode. With a strong current we should not succeed, because under its influence the iron easily becomes passive; in this case, chlorine is disengaged upon the surface of the iron, and oxidizing the carbon already deposited, produces with it a direct combination analogous to hydrochloric acid (?) which the current decomposes, forming a deposit of carbon at the negative pole. In both cases there would be a loss of carbon; in the first, in the form of carbonic oxide or carbonic acid; in the second, in the state of a carburet of hydrogen derived from the carbon and hydrogen arriving together at the negative pole.

"Take a piece of iron, from 10 to 15 grammes, fix it by means of a pinching screw on a platinum point, and plunge it into the dilute acid in such a manner that the point of contact of the iron and its holder be not wet with the hydrochloric acid (otherwise the carbon deposited on the platinum would affect the whole operation), attach the holder to the wire of the positive pole, plunge also the plate of platinum forming the negative electrode into the acid, and by the distance maintained between the two electrodes regulate the intensity of the current, so that only the protochloride and not the perchloride of iron shall be formed. The production of the latter is recognized by the yellow streaks of more concentrated chloride of iron which pass off from the piece of iron. The exterior appearance of the iron changes but little during this operation, because the carbon preserves nearly the form of the iron. When the immersed portion of the cast-iron is dissolved (usually in about 12 hours after immersion), stop the current, separate the compact metallic portion from the carbon mass adhering to it, weigh after desiccation, and ascertain the amount of iron dissolved. In this way the slag is not attacked, but remains in the carbon mass we separate from the undissolved portion of the iron. Collect all on a small filter, heat to redness until all the carbon is burned off, boil the residue with carbonate of soda, to carry off the silica, which is only in the state of simple mixture, heat the residue to redness in a current of hydrogen, and then in a current of dry chlorine, free from contact with the air, treat the mass with a little dilute hydrochloric acid, then again with a boiling solution of carbonate of soda, wash, dry, and weigh. If the quantity of slag be not sufficient for analysis, determine the proportion of silica in it, and calculate the amount of oxygen of the bases combined with silicic acid, according to the approximately known ratio between the oxygen of the acid and that of the bases in analogous slags."

This method is ill-adapted to an industrial laboratory. We are inclined to think that the desired result may be obtained more simply, and with sufficient accuracy, by the following process:—

Dissolve the cast-iron with bromine water, which leaves the slag, carbon, etc., as an insoluble residue; filter and wash this residue, then treat it directly without calcination, with a boiling solution of carbonate of sodium, which dissolves the free silica; filter, wash, and calcine the matter left unacted on by the carbonate of sodium, and in this last residue proportion the silica of the slag in the usual way, having fused it with carbonate of potassium and a little saltpetre.

It seems extremely doubtful whether either of those processes can afford an exact determination of the amount of slag, or "cinder," mechanically involved in cast or in wrought iron. The total quantity in marketable iron is very small, and from the nature of the slags, as usually involved in cast-iron at least, it is scarcely possible but that more or less must not be decomposed

17

by the hydrochloric acid or the bromine, and so rendered soluble. The inquiry is of very small industrial importance in any case. There is much more involved "cinder" from the puddling process in wrought iron bars occasionally than the writer at least has ever remarked in pig-iron, yet Mr. Sorby's microscopic examination of polished specimens have indicated how very minute is the total amount of "cinder" to be found in well-manufactured wrought iron. It is stated in some books of authority, that the fibrous or striated structure developable in wrought iron by the slow surface action of dilute hydrochloric or sulphuric acid (as first pointed out by the late Professor Daniel), is due to the eating out from the mass elongated strips of involved cinder ; and in fact some authors have deemed the "fibre" of a well-rolled bar to be due to the stretching out of portions of the iron by differential movements permitted by the presence of involved, and at the same time liquid, slag or cinder.

These are unquestionably errors, however. The very finest bars, such as a piece of good Lowmoor, or Staffordshire, "rivet iron," and not containing a particle of "cinder," are nevertheless the most perfectly and uniformly fibrous, and when dissected slowly by immersion in acids, are yet acted upon so unequally as to produce sharp fibrous ridges stretching along the piece with deep clefts between. The writer claims to have been the first to point out the true law governing the development of crystalline and fibrous structure in iron in all its various states, in his work "On the Physical Conditions involved in the Materials for the Construction of Artillery," in 1854 They are important to be fully understood by every chemist engaged in iron industry.

[We do not entirely agree with the English editor as to the little practical value of determining separately the silica from the silicon and the cinders, since English rails have been found to contain as much as 5 per cent. of cinder.

Good cast-iron and cast-steel seldom contain cinders inside. The outside deposit on pigs or sandage is generally deducted from the weight of the metal, and if a test is to be made for silicon, the sample should be taken from the inside. In rolled metals the cinder is not regularly distributed, therefore the sample for analysis should be taken from various places of the bar, and especially from sections perpendicular to its axis.

Instead of bromine water, Professor Eggertz employs iodine

or bromine with a certain proportion of water, just enough to dissolve the iodide or bromide of iron formed during the reaction. For instance, 3 grammes of finely divided metal are mixed with 15 grammes of boiled, but cold, water, and 15 grammes of iodine added by small portions at a time. The covered vessel is kept in ice-water, and the contents are stirred now and then carefully, so as to avoid an elevation of temperature, which would result in the formation of oxides and of basic salts.

When the solution is complete, 30 grammes more of boiled and very cold water are added to it, and, after stirring and allowing to settle, the whole is filtered, and the washing is continued with cold water. The filtrate contains iodide of iron and a small proportion of soluble silica (from the silicon), and the filter holds the cinder, the oxide of iron, and the remainder of the silica resulting from the silicon in the metal, plus carbon, etc. We do not calcine the filter and its contents in order to avoid the possibility of a combination of the silica (from silicon) with the cinder. The contents of the unfolded filter are carefully washed and brushed over into a platinum dish, in which the silica is dissolved by a saturated solution of carbonate of sodium mixed with twice its volume of water. After boiling for about one hour the liquor is filtered, and the residue is again treated with a small quantity of alkaline solution, which is also filtered. The solution of silica in carbonate of sodium is acidified by hydrochloric acid, and mixed with the iron solution, and the whole evaporated to dryness on the water-bath. Thirty cubic centimetres of hydrochloric acid are then added, which are also evaporated in order to render the silica insoluble. This substance is determined in the usual manner, and its purity ascertained by evaporation with hydrofluoric acid or fluoride of ammonium, because it may contain a small proportion of vanadic or tungstic acids.

What remains upon the filter is the cinder, carbon, oxide of iron, etc., which are determined by some of the methods indicated in this work.

The operation is conducted in a similar manner when bromine is employed instead of iodine. The proportions are 3 grammes of finely divided metal, 6 cubic centimetres of bromine, and 60 c. c. of very cold water, previously boiled. The vessel is kept in ice-water.

The metal may be dissolved in pieces, instead of in powder or filings, but the operation is much longer, and the surface of the metal ought to be kept clear from carbon by rubbing it with a glass rod.

It is very likely that during this operation, a certain quantity of hydrobromic or hydroiodic acids is formed, but their proportion will be too small to act sensibly on the slag or oxide, if the analysis is performed rapidly, and at a low temperature. It is always very difficult to render the whole of the silica insoluble, especially when it is mixed with a large proportion of iron salt.]

IV.—DETERMINATION OF PHOSPHORUS.

Dissolve 3 to 5 grammes of pulverized cast-iron or iron turnings either in nitric acid or in aqua regia, made with three volumes of hydrochloric acid and one volume of nitric acid; we may also use bromine, but we cannot use hydrochloric acid alone, because it would liberate a part of the phosphorus in the form of *phosphamine* (phosphuret of hydrogen).

Whatever agent be employed, evaporate to dryness to render the silica insoluble, and take up the residue with an aqueous and boiling solution of hydrochloric acid, determining the quantity of phosphoric acid contained in the liquid by one of the methods detailed in a preceding chapter. We can also treat the solution of the iron neutralized with ammonia with the hydrosulphide of ammonium; the iron is precipitated as a sulphide; filter, and in the liquid precipitate phosphoric acid with sulphate of magnesium.

[If the iron is dissolved in concentrated nitric acid, and the solution evaporated to dryness, we may obtain all the phosphoric acid and part only of the iron in solution, by digesting for 12 hours the dried residue in dilute nitric acid. As we get rid in this manner of the greater part of the iron, the subsequent operations are rendered more easy.]

V.—DETERMINATION OF SULPHUR.

First Method: by direct transformation of Sulphur into Sulphuric Acid.

Treat in a covered vessel 3 to 5 grammes of the cast or wrought iron in fragments with fuming nitric acid; as soon as the effervescence diminishes heat to assist the solution, and, when that is completed, add hydrochloric acid and evaporate to dryness. The residue having been treated with hot dilute hydrochloric acid, filter to separate the insoluble matter.

The sulphur changed into sulphuric acid remains in the liquid; precipitate with chloride of barium in the form of the sulphate of that metal. Collect the sulphate, wash, dry it, and weigh after calcination.

[The sulphides of iron are irregularly distributed in the metal, and will be found in greater proportion at the top than at the bottom of a pig, for instance. Therefore, a test sample should be taken from different parts of the piece to be examined.

Nitromuriatic acid does not transform all the sulphur into sulphuric acid, and a certain proportion escapes in the state of hydrosulphuric acid. Professor Eggertz gives the following method: put 5 grammes of finely divided metal and 10 grammes of chlorate of potassium in 200 cubic centimetres of water. Boil, and add 60 c. c. of hydrochloric acid by small quantities at the beginning and more rapidly afterwards. Continue the boiling

17*

and stirring until all the metal is dissolved. Then filter, and precipitate the sulphuric acid with chloride of barium, etc.]

Second Method: by the previous transformation of Sulphur into Hydrosulphuric Acid.

This method is founded on the faculty which sulphurous cast-iron possesses of parting with its sulphur, in the form of hydrosulphuric acid gas, accompanied by the phosphuret and carburets of hydrogen and free hydrogen gases, when attacked by a non-oxidizing acid, such as the hydrochloric. By passing the gases into a metallic solution capable of retaining the sulphur and producing an insoluble sulphide, we can collect the latter, and changing it into sulphuric acid, weigh it in the form of sulphate. The solution of the iron may be made in the apparatus

Fig. 11.

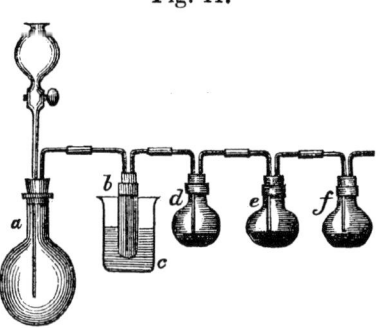

represented in Fig. 11. *a* is a flask of about 300 cubic centimetres in capacity,* carrying a capacious

* All these dimensions are only approximate, and may be much varied.

funnel tube (furnished with a cock), and extending to the bottom of the flask; a gas-tube doubly bent places the flask in communication with a small condenser (formed of a test-tube) two centimetres in diameter and ten centimetres long, containing a very small quantity of water. This condenser is plunged in a vessel filled with cold water. The apparatus is completed by a series of little flasks, *d, e, f,* each 80 cubic centimetres in capacity, half filled with the liquid intended to absorb the hydrosulphuric acid.

Put 3 to 5 grammes of cast-iron in the flask, or more, if the proportion of sulphur be very small; add a small quantity of distilled water; arrange the apparatus, taking care that all the corks are perfectly air-tight, then pour in, gradually, slightly diluted hydrochloric acid through the funnel tube, and regulate the supply, so as to produce the escape of gas bubble by bubble. When the last addition of acid produces no apparent reaction, heat the flask, slightly at first, then progressively, until the liquid commences to boil, and keep it at that point for some time. The hydrochloric acid and steam which are liberated are condensed in *b,* which fills rapidly. When the liberation of gas through the flasks, *d, e,* and *f,* has ceased, stop the operation, which has lasted two to three hours, decant, and treat the precipitate contained in the absorbing flasks. The nature of this precipitate and the treatment to which it is subjected depend upon the reagent employed. Either an alkaline solution of oxide of lead or a neutral solution of nitrate of silver may be used as absorbents.

1st.—*Alkaline Solution of Oxide of Lead.**

The action of hydrosulphuric acid upon this reagent forms sulphide of lead; separate this by filtration on a small filter from the excess of the reagent employed, wash it a little, dry, and fuse it along with the filter it is on, with carbonate of soda and a little saltpetre. The melted mass is treated with water, the solution is subjected to a current of carbonic anhydride, in order to precipitate any traces of dissolved lead, and then separated from the insoluble matter by filtration. Then acidulate the liquid with hydrochloric acid, and precipitate the sulphuric acid with chloride of barium, using the precautions indicated already.

2d.—*Solution of Nitrate of Silver.*

Introduce into each of the little flasks, *d* and *e*, forty cubic centimetres of a solution of nitrate of silver, containing one part of the nitrate to twenty parts of water. In flask *f*, twenty cubic centimetres of this solution are placed, adding thereto an equal volume of water.

By the action on the nitrate of silver of the gases liberated by treating cast-iron with hydrochloric

---

* The alkaline solution of oxide of lead is prepared by adding caustic potassa in small pieces, or in strong solution, to a solution of the nitrate or acetate of lead, in quantity more than sufficient to completely redissolve the precipitate which forms at first. The solution is of a suitable concentration if it contain 25 to 30 milligrammes of lead for each cubic centimetre of liquid.

acid a black precipitate is formed, principally composed of the sulphuret and phosphuret of silver. There is a slight disturbance in all the absorbing flasks, produced by the reduction of a small quantity of silver by the evolved hydrogen. Pour the contents of all the little flasks on the same filter, beginning with the one farthest from flask a, wash them out once with distilled water, and use this water to wash the precipitate. After this imperfect, but sufficient washing, pour some water and a little bromine into flask f; shake and warm slightly to dissolve the bromine, then plunge the ends of the tubes from d, e, and f, into the liquid. The black matter adhering to them changes rapidly into yellow bromide of silver; wash now the tubes with distilled water, and then pour the bromine water from f into e; when the small quantity of black precipitate in it has also been changed into bromide of silver, pour the liquid into the third flask, and from that, when it has acted upon the matter adhering thereto, into a matrass of 250 or 300 cubic cent., wash all the absorbing flasks, and add the water to the first liquid, so as to unite the whole.

Add to this a new quantity of bromine, place the funnel containing the precipitate of the sulphide over the matrass, pierce the filter, and, aided by a little water, cause the greater portion of the precipitate to fall into the matrass, into which finally put the filter itself. Heat slightly to be sure that the action is complete, and to expel the excess of bromine, which should until then be carefully preserved in the liquid.

The reaction between the sulphuret of silver, the water, and the bromine, may be represented by the formula—

$$Ag^2S + 8Br + 4H^2O = H^2SO^4 + 2AgBr + 6HBr.$$

The phosphuret of silver is transformed by an analogous reaction into bromide of silver and phosphoric acid. Separate the bromide of silver by filtration; the acid and hot liquid is quite free from fixed matters and in good condition for the precipitation of the sulphuric acid with chloride of barium; be careful to use only a slight excess of this reagent, which can easily be effected by employing a standardized solution.

[Instead of a neutral solution of nitrate of silver, made by dissolving the dry crystals of nitrate in water, many chemists employ ammoniacal solutions of nitrate of silver, or of chloride of copper, or of chloride of zinc.

A rapid colorimetric process for the determination of sulphur in iron, and which gives approximate, but useful, results when the proportions are small, has been proposed by Professor Eggertz. A bright plate of an alloy of silver (75 parts) and copper (25 parts), is suspended in a closed tube, on the bottom of which there is the sample of iron (0.1 gramme) to be tested, and enough diluted sulphuric acid to dissolve it. The sulphur, if any, is transformed into hydrosulphuric acid, which colors the silver alloy a shade corresponding to a given proportion of sulphur, determined by previous experiments. The alloy becomes yellow, brown, and steel-blue, with all intermediary shades. Steel-blue corresponds to 0.2 per cent. of sulphur in iron. The determination of the shades is assisted by a comparison with various alloys accompanying the instrument. More detailed accounts of the process are to be found in Percy's Metallurgy of Iron and Steel, and Gruner's Manufacture of Steel.]

VI.—DETERMINATION OF CARBON.

Of all the elements which enter into the composition of the industrial products with which we are here engaged, carbon is, after the iron itself, the most important. It is generally admitted that it may exist in *cast-iron* in two forms—1st, in the state of crystallized carbon or *graphite;* and 2d, as *dissolved* or *combined carbon*. Iron and steel contain it in the latter condition only. The weight of the *whole* of the contained carbon is generally ascertained, but it may also be desirable to fix the proportions of the carbon as found in each form. In the analyses in which we are engaged it is advisable always to pursue the same method of operation; for in general different methods will give different results. By always operating in the same way we can at least insure comparable results, and that is the essential point.

A.—DETERMINATION OF THE TOTAL AMOUNT OF CARBON.

It is generally admitted that the determination by direct weighing of the carbon separated from cast-iron gives rise to error, as the separated carbon invariably contains hydrogen and even oxygen in combination. The only method leading to exact results is to burn the carbon, and to ascertain the amount in the form of carbonic anhydride, condensed in a solution of potassa, contained in an absorbing bulb apparatus, previously weighed. The increase of weight gives what we require. The transformation of carbon into carbonic anhydride can be done directly by submitting cast-iron to the action of certain oxidizers, or by first liberating the carbon, and then burning it.

1st.—*Determination of Carbon in Cast-Iron by direct Combustion.*

The most simple mode of effecting direct combustion is that proposed by Wöhler; that chemist burns finely pulverized cast-iron in a current of oxygen. It is obvious that cast-iron, even when pulverized, forming a more or less compact mass, oxidizes with difficulty. To render the mass more permeable to the gas, Rose adds the oxide of copper.

Instead of employing free oxygen as an oxidizer, we may burn cast-iron by heating it with a substance that will yield its oxygen to it. Kuddernatsch employs oxide of copper only. We can hardly believe that complete oxidation can be obtained by this method. Regnault and Bromeis proposed a mixture of chromate of lead and chlorate of potassium; but Kuddernatsch, having remarked that this mixture liberates chlorine when exposed to the action of heat, it is preferable to employ a mixture of chromate of lead (ten parts), and bichromate of potassium (one part), as indicated by Mayer. The combustion of the cast-iron is effected in apparatus and by methods quite the same as an organic analysis—viz., in a tube of refractory glass, heated either with charcoal in Liebig's iron-plate furnace, or, which is preferable, by means of coal gas in a special apparatus. The apparatus and the manipulations being nearly the same as those used for the combustion of carbon liberated from cast-iron, we refer to that article for detailed description.

2d.—*Determination of Carbon after liberation from the Iron.*

a. *Previous liberation of the Carbon.*

First Method: by Bromine.

Operate with 5 to 10 grammes of cast-iron in fine granules placed in the bottom of a narrow vessel containing water, into which pour afterwards sufficient bromine to cover the iron and prevent it coming in contact with the water. Let the bromine act at the ordinary temperature; stir the iron in the bromine gently from time to time. Decant the water containing bromide of iron now and then, replacing it with fresh water, and make a new addition of bromine, if necessary.

When we consider the solution complete, which in certain cases is not for some days, carefully decant the water and expel the excess of bromine by gentle heat. Add hot water and a little hydrochloric acid.

The weight of the carbon is ascertained according to one or other of the methods indicated further on.

Second Method: by Chloride of Copper.

Operate with 5 to 10 grammes of cast-iron, reduced to granules smaller as the iron is whiter and harder,

18

and at a temperature not exceeding 50° Cent.,
submit it to the action of a solution of chloride of
copper as neutral as possible, and containing at least
a quantity of copper equal to one and a half that of
the cast-iron employed. The reaction between the
iron and the chloride is represented by the following :—

$$Fe^2 + 2CuCl^2 = Fe^2Cl^4 + Cu^2.$$

The iron takes the place of the copper, abandoning
its carbon, which is mixed with the metallic copper
precipitated. The action is assisted by frequent
stirring, and when over (which is known by the
absence of hard grains in the undissolved residuum
when felt with a glass rod) decant the liquid into a
second vessel, and treat the mixture of copper and
carbon with a fresh portion of solution of chloride of
copper, mixed this time with hydrochloric acid. By
this means the copper is dissolved, and in the state
of chloride of copper is soluble in hydrochloric acid—

$$Cu + CuCl^2 = Cu^2Cl^2.$$

The carbon remains along with the silica, etc.; the
mode of determining its weight is given further on.

It has been proposed to replace the pure chloride
of copper by a mixture of this salt and chloride
of sodium. The object of the latter is to render
the cuprous chloride soluble by forming with it a
double salt, and thus avoiding the employment of
hydrochloric acid. We may also substitute for the
chloride of copper a mixture composed of the sulphate
of this metal and chloride of sodium. The solution
employed in this case is formed of twenty parts of

crystallized sulphate of copper and twenty parts of chloride of sodium with 100 parts of water.

Third Method: by Bichloride of Mercury.

Triturate in a porcelain mortar one part of cast-iron in powder or filings, with fifteen to twenty parts of bichloride of mercury, and a sufficient quantity of water to form a thin paste; the chloride of mercury acts upon the iron, and protochlorides of iron and mercury are formed—

$$Fe^2 + 4HgCl^2 = Fe^2Cl^4 + 2Hg^2Cl^2—$$

the carbon remaining intact. The reaction has terminated when no more hard grains remain under the pestle, which takes place after about half an hour. Dilute the paste with 200 to 250 cubic centimetres of hydrochloric acid, and leave all in the oven for about an hour at a temperature of 60° to 80° Cent. The protochloride of iron and the mercuric chloride not decomposed are dissolved; the protochloride of mercury remains with the carbon and silica as an insoluble residuum. This is thrown on a filter and washed with warm water, dried, separated from the filter, and placed in a platinum boat. This is put into a tube of refractory glass and heated there by means of a gas or charcoal furnace in a current of pure, dry hydrogen. The whole of the protochloride of mercury volatilizes and condenses either in the colder parts of the tube or in a flask fitted to it for the purpose of collecting the mercurial vapors, which are dangerous to inhale. Allow the platinum boat to cool in the current of hydrogen, then weigh it, having first inclosed it in a small tube of thin glass, fitted with a cork, to prevent the absorption of moisture by the carbon. After weighing, the boat is heated first in contact with the air or a current of oxygen, in order to burn off the carbon, then again in a current of hydrogen, in order to bring back the residuum of this combustion to its condition when first weighed. By weighing a second time we obtain the weight of the carbon by the difference. We might also determine the weight of the carbon by the quantity of carbonic anhydride it produces as hereafter stated.

Fourth Method: by Galvanic Current.

This method has the advantage of the others in not requiring the mechanical division of the specimen. We have already detailed the mode of operating according to Frésénius in reference to the determination of silicon, which is also here to be employed. The carbon residue is separated from the liquid, and the weight of the carbon it contains is determined by one of the methods here given.

[Wöhler employs for the previous liberation of the carbon another method, by which chlorine gas is made to pass upon the metal kept in a porcelain boat, and heated to a dull redness in a glass or porcelain tube. This process is rapid, and presents the advantage of not requiring that the metal should be finely divided. When the chloride of iron has been volatilized, the carbon remains in the boat and is cooled there in an atmosphere of chlorine. This gas should be entirely free from hydrochloric acid, in order to avoid a loss of carbon in the shape of escaping hydrocarbons. There may, possibly, be some light particles of carbon carried away from the boat by the gases, but we believe that the loss from that cause is very small.]

b. *Determination of the weight of the Carbon eliminated.*

Whatever method we employ to obtain the liberation of the carbon, it is always accompanied with foreign bodies. We have already stated that the only way *accurately* to determine the weight of the carbon is to transform it into carbonic anhydride, and to weigh it in that state. Still as we may not have the apparatus required for this always at our disposal, we should notice two other methods giving approximate results.

First Method.

Throw the carbon residuum separated from the cast-iron on a weighed filter, having submitted it to the action of dilute hydrochloric acid; wash it well, determine the weight after desiccation at 100° Cent., and then calcine. The carbon is burnt off, and nothing is left but the fixed matter which had been mixed with it. Weigh the residuum, and regard the difference between the weight now obtained and that furnished by the preceding method as representing that of the carbon. This result is not exact: the error arises from the combined carbon that has been separated from the cast-iron containing hydrogen and

18*

oxygen, and from the silica mixed with it not being completely dehydrated at 100° Cent.

[Silica obtained by precipitation and dried at 100° C., contains very nearly 94 per cent. of perfectly dry, or calcined, silica.]

Second Method.

The filter upon which the carbon is thrown is placed with its contents in a small porcelain crucible, covered closely, and then placed in a larger crucible, and covered completely with small pieces of wood charcoal. The larger crucible is provided with a cover, and the whole is heated red hot in a muffle furnace until the filter is completely carbonized. Cool before opening, and then ascertain the weight of the matter contained in the interior crucible. This weight includes that of the carbonaceous residue of the filter, that of the carbon separated from the cast-iron, and that of the fixed matter mixed with it. We obtain the weight of the latter by burning off the carbon of the cast-iron and that of the filter. We ascertain the weight of the substance left after the calcination of the filter in a close vessel, and thus the difference gives the weight of the carbon. To obtain the weight of the filter, make a certain number of the same size, and equal in weight, of paper of the same thickness and grain; calcine several of these filters in a crucible from which the air is well excluded; weigh the carbon arising from the carbonization of the paper. By conducting all the operations in exactly the same manner, we obtain, after some trials, weights perceptibly equal for all filters of the same size.*

Third Method: by Combustion.

The liquid obtained by attacking the cast-iron is filtered on a plug of asbestos† or a sponge of platinum, placed in the bottom of a tube tapering at the end, or in a small funnel. Having washed the carbon residuum, first with hydrochloric acid, then with pure water, until the acid be completely got rid of, dry it at 100° Cent., or at a temperature a little higher, in the same tube in which it was collected by filtration. Mix it with peroxide of copper, and effect the combustion by means of its oxygen. The apparatus applicable to this purpose is represented by fig. 12. a, b, c, is a tube of refractory glass, from 60 to 70 centimetres long, and 12 to 15 millimetres interior diameter, filled from b to c with oxide of copper in grains, retained between two plugs of asbestos or of fine copper wire; d is a tube filled with chloride of calcium; e, an apparatus filled to about a third with concentrated solution of caustic potassa (1 part of caustic potassa and 2 parts water); and f, a small tube half filled with potassa in small fragments, and half with chloride of calcium; g is a gasometer filled with oxygen; h, a tube of U form, filled with sodic lime, in order to retain the carbonic

* Rivot's "Docimasie," t. iii. p. 522.

† The asbestos should be purified with boiling hydrochloric cid, and then calcined.

anhydride which the oxygen often contains, in
consequence of the presence of small quantities of

Fig. 12.

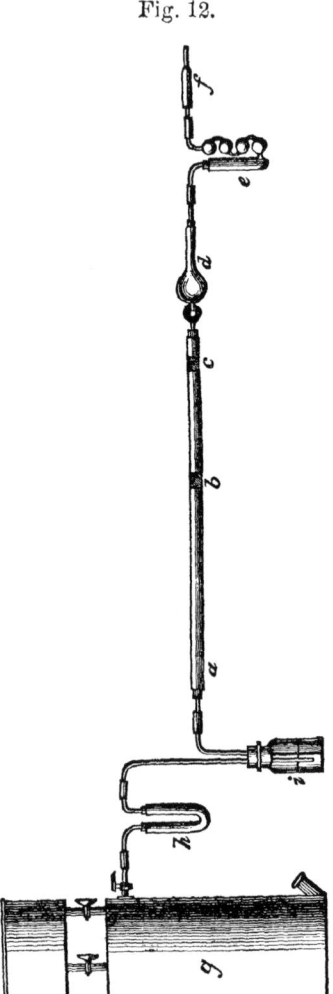

organic matter in the oxide of manganese or other bodies from which it is prepared; lastly, i is a washing-flask containing sulphuric acid, and intended to dry the gas, and to indicate the rapidity of its liberation. The different parts of the apparatus are united with well-fitted corks or caoutchouc tubes.

Put a small quantity of powdered oxide of copper in a very dry porcelain mortar, and spread it with the pestle; by a copper or platinum wire, or a glass rod, cause the carbon, and the plug which held it in the tapered or funnel-shaped tube before described, to fall into the mortar; then pour a little oxide of copper into the tube, and, turning it round, detach any carbon which may be adherent to it by the friction of the oxide. Add this oxide to that in the mortar; mix all well together, and introduce it, by the extremity a, into the part $a\,b$ of the combustion tube. Clear perfectly the pestle and mortar by means of some more oxide of copper, which may be first passed through the taper-tube funnel to carry off the last traces, if any, of the carbon, and put this also into $a\,b$. Put together the apparatus as shown in the figure, having accurately weighed the vessel containing the caustic potassa solution, and the tube f, together; then begin to heat the oxide of copper contained from b to c, taking care to protect the cork with an iron screen. When this first part has been heated red, heat the remainder of the tube progressively, advancing from b to a. At the same time cause the oxygen to pass, bubble by bubble, through the apparatus.

Under the combined influence of the oxide of

copper and oxygen the carbon burns, and is changed into carbonic anhydride; the gases escape through tube *d*, where they are dried, and through the vessel containing the caustic potassa solution, where the carbonic anhydride is absorbed and retained.

The excess of oxygen only passes through tube *f*, where it is deprived of the water it had carried off from the solution of potassa, and from thence it passes into the air. When the combustion is terminated, detach the tube *h* from the gasometer, and let a current of air pass through the apparatus for five minutes, by means of an aspirator fitted at *f*. Detach the tubes *e* and *f* from the tube *d*, and weigh again; the increase of weight represents the amount of carbonic anhydride absorbed, and from this weight we calculate the weight of carbon in the cast-iron submitted to analysis.

B.—DETERMINATION OF THE UNCOMBINED CARBON, OR GRAPHITE.

Treat from 5 to 20 grammes of cast-iron according as it contains more or less graphite, with hydrochloric acid diluted with $\frac{1}{2}$ to 1 volume of water. Heat slightly to assist the solution, and when that is effected, which we ascertain by the liberation of gas ceasing, dilute with a little water and collect the residuum, pouring the liquid on a weighed filter Dry at 100° Cent., and weigh. Calcine the residuum; the graphite and the filter are burnt together; we ascertain the weight of the fixed matter remaining, and, by the difference, we learn the weight of the graphite. This method is not free from error. To obtain accurate results, we should filter the solution obtained with the aid of hydrochloric acid, through a platinum sponge or asbestos plug, as described in the paragraph relative to the determination of the total amount of carbon, and wash, first with boiling water, then with a solution of potassa; remove the last traces of this by washing with alcohol, and, finally, wash with ether. Dry the residuum, and effect the combustion as already directed.

C.—DETERMINATION OF CARBON OF COMBINATION.

First Method: by Difference.

Ascertain separately the total amount of carbon and the amount of graphite, according to the methods just indicated. Subtracting the second from the first, the difference gives the amount of carbon of combination.

Second Method: by Colorimetric Comparison.

This process is founded on the property possessed by cast-iron and steel, when dissolved in nitric acid, of producing a brown color, *deeper* in proportion as the amount of their carbon in combination is greater. The pernitrate of iron which is formed does not communicate any perceptible tint to the liquid; and the graphite, if the iron contain it, remains intact. Like all colorimetric processes, the question is determined by comparison with a typical liquid, or visual standard.

In the case before us, this standard liquid is prepared in the following manner: Dissolve one to two decigrammes of cast-steel, containing a known amount of carbon, in 1.5 to 5 cubic centimetres of nitric acid, of density 1.20, and free from chlorine. This last condition is important, because the chlorine would form perchloride of iron, the acid solution of which possesses itself a decided coloring power. The amount of acid should equal that required for the complete solution of the specimen to be tested, and proportionate to the amount of carbon it contains.

The steel employed for the standard liquid should be in powder: pulverize a sufficient quantity, and pass it through a metallic sieve, the apertures of which have a maximum diameter of $\frac{4}{10}$ of a millimetre. Ascertain the total amount* of carbon accurately, by one or other of the methods previously detailed, and secure the metallic particles from damp and air, which would cause their oxidation.

The solution is made in a glass tube, 10 to 15 centimetres long, and about 1 centimetre in diameter; heat to 80° Cent. in a water-bath, regulating the temperature by the thermometer. The blackish flakes formed at first disappear upon the liberation of gas. After two or three hours the solution is complete.

To assay by this method iron and steel, begin by reducing them to powder, and passing them through the sieve; then weigh a quantity equal to that of the typical or standard steel, by means of which the

---

* All the carbon in steel is in combination.

liquid for comparison has been prepared; treat it in a glass tube, similar to that used for the preparation of the solution of the typical steel, for the same length of time, and with the same quantity of acid, and exactly in the same way. The black particles which the liquid may still contain, after two or three hours' reaction, can only be attributed to graphite or slag: we can ascertain this by heating them with some drops of nitric acid, having first decanted the liquid in which they were deposited; in this case there should be no liberation of gas.

When the solution of the typical or standard steel and of the specimens for assay are completed (for the sake of greater accuracy, several repetitive assays should be made at the same time), bring back the liquids to the ordinary temperature, by plunging the tubes containing them into cold water, because the coloring is more intense while hot than cold, and we can only compare the shades at the same temperature. To make this comparison, take two glass tubes as nearly alike as possible; in one put the typical solution, and dilute with water, so that every cubic centimetre corresponds to $\frac{1}{10}$th of a milligramme of carbon. It is an advantage to use a tube which is graduated in accordance with the volume required, and which will remain always the same for the same steel. Into the second tube pour the solution of the specimen to be assayed; dilute it gradually with water until the two tubes viewed by transmitted light or by reflected light, with a sheet of white paper behind them, possess exactly the same tint. Then measure the volume of the liquid, by

19

pouring it into a graduated tube or jar, marked in cubic centimetres. The number of the latter, if the operation be performed as indicated, expresses the number of tenths of a milligramme of carbon contained in the weight of the steel or cast-iron assayed, and enables us to calculate it.

In assaying iron or steel poor in carbon, we must use a typical solution, each cubic centimetre of which corresponds to five hundreths of a milligramme of carbon only. For cast-iron with a large proportion of carbon, the colorimetric process is wanting in accuracy; but for testing steel, especially Bessemer steel, it gives excellent results, admitting of fair comparison. It is largely employed, on account of its simplicity and rapidity, in steel manufactories, when materials have to be compared of apparently similar composition and constitution; but we cannot depend upon the results of a comparison between steel and cast-iron, especially if the latter contain large quantities of sulphur, phosphorus, and silicon.

[This colorimetric assay is really useful for those compounds of iron which contain a low percentage of combined carbon, with little or no graphite, and which take so long and are so difficult to analyze with accuracy by the gravimetric processes. Another condition is that the visual or colorimetric standard should be prepared from a standard metal having great analogy, in the metals used and the mode of manufacture, with the sample to be examined. Errors will result from the indiscriminate use of the same colorimetric standard for the examination of steels and irons coming from different localities, and prepared with different ores and modes of manufacture. The shade of the colored nitric solution is influenced not only by the combined carbon, but also by other substances which may exist in the metal. In other words, the same coloring standard, prepared from a standard metal, may be employed for the comparison of other metals prepared in the same works with similar materials and by a similar mode of manufacture; but in the case of other steels, different in materials and mode of preparation, it will be necessary to make a new colorimetric standard from a standard metal, analyzed by gravimetric process, and similar to the new kinds of steel under examination. A more extended description of this process will be found in Percy's Metallurgy of Iron, and in Gruner's Manufacture of Steel.]

GENERAL ANALYSIS OF CAST-IRON.

We may rapidly determine almost all the elements of cast-iron by the following process: Determine the amount of sulphur by treating 5 to 10 grammes with hydrochloric acid, and by causing the gases to pass into a solution of nitrate of silver (as previously described); and while treating the precipitate of sulphide and phosphide of silver with bromine, filter the hydrochloric solution on a weighed filter, and determine the amount of graphite contained in the residuum as already directed. When this has been calcined, fuse it with sodic carbonate of potassium, treat the melted mass with hydrochloric acid, add the solution thus obtained to the liquid proceeding from the first filtration, and evaporate all to dryness. Dissolve again in diluted hydrochloric acid; the residue is the silica corresponding to the total amount of silicon. In the liquid separated from the bromide of silver the sulphur is now changed into sulphuric acid, precipitate it with chloride of barium; the liquid separated by filtration from the sulphate of barium is deprived of baryta by means of sulphuric acid, and then united to that left after the filtration from the silica. In this manner we obtain a solution containing all the iron, manganese, and phosphoric acid: dilute with water to the volume of a litre, then take known parts thereof and determine in them *volumetrically* the iron, and the manganese and phosphorus according to the methods indicated in the preceding pages. The determination of the carbon in combination by the colorimetric process completes this simple and rapid analysis.

There is no characteristic more remarkable in the chemistry of iron than the wide range of its affinities for foreign elements, the tenacity with which it retains in combination the last traces of several of these bodies, and the remarkable influence which those minute proportions of foreign elements exercise upon the physical properties of the metal. And amongst the bodies thus combined in small proportion with iron, as powerfully affecting those physical properties, carbon stands foremost; hence the exact determination of this element, both as to its amount and as to its state, *i. e.*, as combined or as diffused—constitutes a problem of almost unique importance in sidero-technic analysis.

The methods employed or proposed, numerous as they are, and emanating from some of the greatest amongst chemists, still leave much to be desired.

Before adding a remark or two to the instructions given in the text of Part VI., it may be well to observe that the statement (foot note to sec. C), that all the carbon in steel is found in a state of combination, is not quite or universally true. Karsten's views, long since promulgated as to the states in which carbon is found in cast-iron, in wrought iron, and in steel, though by a few persons ineffectually impugned, may be held now as substantially admitted. In commercial cast-iron, then, we have much graphite present, and more or less combined carbon ; in wrought iron very little carbon in either state, and that which exists chiefly in combination, and in steel and *spiegel eisen*, as well as in perfectly "chilled" and white cast-irons, *almost* the whole of the carbon present is chemically combined. But even in cast-steel some traces of graphite are to be found after solution, and more or less graphite is nearly always present in wrought iron. We may, therefore, deceive ourselves if we assume that because we operate upon a material, commercially called steel, we have solely to do with combined carbon.

The scales of graphite liberated from wrought iron and steel containing carbon in this condition are extremely minute ; so much so as to remain long suspended in solution, and only evident by a slight want of pellucidity in the solution, or by a peculiarity in its tinge of color, and in this state may add an element of uncertainty to the colorimetric indications of Eggertz's method, one in which the writer must confess he has but little confidence as a quantitatively exact process, subject as it is, like all other analogous methods, to the uncertainties introduced by the want of an invariable standard of color, and by the different way in which different eyes regard the same color, or the same eye regards it at different times. The statements of W. D. Herman as to this process ("Journ. Chem. Soc.," vol. viii., p. 375) should be consulted. The normal or standard solution of steel for Eggertz's method cannot be kept unaltered in color for many hours. It has, therefore, been proposed to form a scale of color by placing solutions of caramel (burnt sugar), or of roasted coffee, in a known series of proportions, as to the water and alcohol, in tubes, wherewith the steel solutions of the assay shall be compared. But are these, though sealed up, invariable in color after more or less exposure to light ?

With respect to the separation of the graphite from the carbon of combination by the assumed insolubility of the former, after both have been liberated from the iron, it may be observed that the researches of Brodie ("Ann. de Chim.," 3 ser. t. xiv., and "Journ. Chem. Soc.," xii.) point to the probability that graphite is by no means always that stable body which is here assumed, so that it is quite possible that under the influence of the more powerful oxydants used to liberate the total carbon, some of the graphite may change its form, if not even be brought into solution or driven off.

Again, the writer has observed some facts which seem to indicate that, as the state of the carbon in iron after it has been solidified from fusion depends in a great degree upon the rate at which it has cooled, so when brought into solution its state in some degree depends, if not upon the solvent, at least upon the rate at which the solution has been effected; in fact, if the solution be *slow enough* there is reason to suppose that more or less of the carbon of combination passes in the act of its liberation into the condition of graphite.

Thus white chilled cast-iron, still more, fine close-grained cast-steel, which if dissolved rapidly, leave almost insensible traces of graphite; if *very* gradually dissolved, as when acted upon by air and sea-water, or by *very* dilute hydrochloric acid, or very dilute solutions of chlorides, bromides, or iodides, leave a pseudomorphic mass, consisting mainly (when dry) of a silvery-gray powder, which reveals itself under the microscope, as consisting chiefly of excessively minute particles of crystallized graphite.

It is thus not absolutely certain, that in liberating the total carbon, by breaking up the iron with a solvent, we always eliminate this in the same proportions as to combined and graphitic carbon in which it has existed in the metal itself.

It is true that in the industrial laboratory comparative results are those most usually demanded. There is, therefore, from the above the greater reason why we should always effect the liberation of the carbon by the same solvents and at the same rate.

As respects the various methods of liberating the total carbon, the writer has not found that by chloride of copper satisfactory, but when employed it is best not to attempt to isolate the carbon directly from the metallic sponge of copper, but, after washing and drying, to burn the whole, mixed with peroxide of copper, and determine the total carbon or carbonic acid. The writer has had no experience with bichloride of mercury as a solvent, and the production at different stages of the insoluble protochloride and of metallic mercury presents some difficulties. M. Boussingault has, however, produced a most able memoir upon it ("Ann. de Chim.," 4th ser., t. xix. 1870), in which ample details are given as to its use, and in which he expresses his confidence in its value. It deserves the study of the iron analyst. Bromine and iodine both act well upon the metal, the former the more energetically, but with a little annoyance due to the irritating vapors which it is impossible to prevent. It is worthy of attention whether fluorine, applied in some form analogous to Brunner's process for action on silicious minerals, might not prove superior to any other agent in rapidity of action and facility of application.

It is not unlikely that fused chloride of silver, reduced to fragments and mixed with those of the iron, and placed beneath water slightly acidulated with hydrochloric acid, or containing a little chloride of sodium, might be advantageously employed to liberate the total carbon; and that the amount of the latter might from the fixity of silver in the fire, be inferred directly from the weight of the reduced silver, after its calcination and exposure to sufficient heat upon a cupel for its fusion to a button; in this case the sulphur, if any be contained in the iron, must be separately determined.

Combustion with oxydants appears beyond question the best method of determining the total carbon; and whether this process be employed upon the iron itself, or upon the carbonaceous residue liberated previously from it by one method or another, must always remain a matter of judgment for the analyst who has the particular sample of iron before him. The question turns mainly on this: If the iron be rich in carbons, moderately soft and friable, and thus capable of being readily brought into a *very* finely-divided state, direct combustion is capable of giving the more exact results, and more rapidly; but if the carbon be in quantity small, be chiefly in combination, and the metal, like cast-steel, *spiegel eisen*, chilled or refinery iron, etc., be excessively hard, less precise results will be procured by direct combustion, and that even though the metal shall have been laboriously brought to a state of fine division; and previous liberation of the carbon will be best.

The writer had occasion some years ago to make a large number of determinations of carbon in irons by combustion, and having had to form his own experience, may note a few points here for the benefit of others. Nothing seems easier, when reading the directions given by authors, than to make a good assay of iron or carbon by combustion, yet very few analytical operations are in reality more difficult or more uncertain until after a good deal of painful experience in failures has made the operator acquainted with the precise march proper for the operation. First, then, as to the iron itself: Its state of subdivision cannot be too carefully procured. The hardest steels, or "chilled" iron, can be brought to any extremely divided state—when the sample is large enough to admit of it—by turning off in the lathe, from a cylindrical bar, very fine shavings, with a crystal of diamond set as a turning tool. In this way a sort of curled steel or iron *wool* can be readily and pretty rapidly obtained, so fine that a little put into the flame of a spirit lamp instantly flashes off, and leaves nothing but oxide. It requires a little time to break it up short, when it is at once fitted for combustion.

With wrought iron this may be practised also, and the turning tool may be of steel ; but *filing* even "with a new file previously deprived of adherent oil," as directed in some works, is to be avoided. The number of steel teeth broken out of the file after even an hour or two's work, is greater than might be imagined ; and filings even when pretty fine, are lumpy little fragments presenting small surfaces in relation to their masses. For cast-iron, turning or planing off, by a tool cutting finely, and not taking too gross a cut, and the tool bent abruptly close to the cutting edge, so as to curl up and fracture in as many places as possible the cuttings, does best. These cuttings can then be pulverized further in a large-sized cylindrical steel mortar.

The writer has tried with cast-iron the plan of levigating those fragments by grinding them up like paint upon a painter's "flag and muller" of a large size, along with oil of turpentine, and removing this by digestion with caustic alkali, and washing with alcohol, or even water, and drying out of contact with air, and with the result of procuring extreme subdivision ; but he is not absolutely certain that no trace of the oil of turpentine is absorbed by the iron particles, so as not to be removable, though he has had no proof either that it is so.

However effected fineness of division is the *sine quâ non* of accurate determination of carbon by combustion.

Then, as to the apparatus, he admits a preference for the old form of organic analysis apparatus as devised originally by Liebig, and for his form of potassa absorbent apparatus (the *kali apparat* of Giessen), consisting of a triple bulb, etc. Charcoal, burnt in the Liebig sheet-iron trough or furnace, he has found more manageable than a row of gas flames, with their attendant troublesome brick or earthen heat confiners, etc. But the sheet iron trough requires to be made more capacious than for usual organic analysis, in order to hold more charcoal and enable the temperature towards the end of the process to be pushed to the highest point the combustion tube will bear. As regards the oxydants, he has never succeeded satisfactorily without employing a current of gaseous oxygen. Aided by that, when applied at the right time and at the right temperature of the combustion tube, either peroxide of copper or chromate of lead may be employed, but the last is decidedly preferable. It is best in the hard condition, procured by fusing in close vessels, and subsequent grinding up finely. In this state it does not agglutinate so soon, or at so low a temperature as the unfused chromate.

The mixture of chromate of lead and chlorate of potassium, he has never been able to use advantageously, for independently of disengagements of chlorine, these mixed salts agglutinate and fuse, and act upon the glass combustion tube much more rapidly than the chromate alone. Towards the end of the process it is advisable to push the temperature to the highest point the glass tube can bear without melting, sagging, or blowing out. Porcelain tubes have been suggested, but the writer found their use difficult in several ways, and more liable to accident by rapid changes of temperature than glass, besides their opacity preventing any visual knowledge of what is going on inside. Black or green British bottle glass was tried, but though rather more refractory than German glass, it has the disadvantage of passing at once almost from solidity to fusion with but little intermediate stages of viscosity.

The wrapping round the combustion tube with a strip of thin sheet-iron is not so good, as merely supporting the lower side of the tube by a semi-cylindrical or trough-shaped guard of thicker sheet-iron. The writer tried with the combustion tubes a continuous coating of copper electrotyped upon them, over a surface thinly varnished with a solution of phosphorus in benzole, with a very little Canada balsam—a method generally applicable to chemical glass vessels, of which he published an account several years ago, but with no great advantage, owing to the fusing point of copper being as low as it is. Now, however, that a like deposited coating of nickel can be readily obtained by Gaiffe's method, a firm and sufficiently infusible jacketing to the combustion tube is obtainable, but the tube is rendered opaque.

In the conduct of the combustion the main points to be observed are, beginning the heating of the combustion tube at the proper end, turn on the current of oxygen gas at once, let it be *very slow,* only bubble by bubble through the potassa bulbs. Keep the temperature of the whole apparatus at not more than a low red heat until the whole length of the tube is thus heated, and until a pretty large volume of oxygen has been drawn through. Now raise the temperature a little, to about a full red, and continue the passage of the oxygen so for about an hour (more if the iron be very hard, or its subdivision defective), and finally raise the temperature to as high a point as the tube will bear, and keep it so, with the oxygen current still slowly passing, until upon shutting it off for a moment or two we see that bubbling through the potassa bulb ceases also. After a little longer the operation may be viewed as ended.

For further remarks containing very masterly instructions for this process, as well as for all that relates to the assay and analysis of iron, the reader is referred to the grand work of M. Rivot ("Docimasie: Traité d'Analyse des Substances Minerales," in four volumes; Paris, 1864), and for a well-collected mass of information on the estimation of carbon and other constituents in iron, to Mr. Crooke's "Select Methods of Chemical Analysis," pp. 67–147.

PART VII.

ASSAY OF FUELS.

THE results of the calorimetric assays to which we can submit fuels are not always sufficient criteria of their industrial value; nevertheless they furnish some hints which can be made use of advantageously. We can only learn absolutely the value of a fuel by observing how it behaves in the blast furnace or grate of a draft furnace wherein it is burnt; but such experiments demand much time, and are expensive. Chemical analysis or simple laboratory assays, when united with experienced observation of external properties, often obviate the necessity of employing those on the great scale. There is seldom occasion for a complete analysis of a fuel; it is sufficient generally to determine the proportion of water, ashes, and sulphur which it may contain, by the methods we are about to describe.

I.—DETERMINATION OF WATER.

The amount of hygroscopic water contained in a fuel is ascertained in exactly the same manner as that before indicated for the same when contained in ores.

II.—DETERMINATION OF THE ASH.

Weigh 1 to 3 grammes of the fuel in a broad and flat platinum capsule, or, still better, in a small tray of that metal, made for the purpose, with a flat bottom and slightly raised edges; spread the specimen to be assayed in a thin layer thereon, and place it in the muffle, submitting it to a gradually increasing temperature, in order to guard against any loss arising from projection by decrepitation, etc. As soon as the combustion of the coal appears complete, withdraw, very gently, the capsule or tray from the muffle, and weigh when cooled under the drying-bell. Heat the residuum again until two consecutive weighings are identical. If we desire to ascertain the composition of the ashes, collect the residue left after the incineration of several specimens; the quantity obtained must be sufficient to permit of a complete analysis, which is performed nearly as that of an iron ore; the constituents of the ashes of most coals being chiefly iron, silica, alumina, lime, magnesia, and traces of alkalies.

III.—DETERMINATION OF THE SULPHUR.

Transform the sulphur into sulphuric acid, by treating the fuel reduced to very fine powder, with fuming nitric acid. The details of the operation are exactly the same as before given with respect to an ore. Bromine may be used with advantage in place of nitric acid.

IV.—CALORIFIC POWER—HEATING VALUE OF FUEL.

The only method to ascertain the amount of calorific power sufficiently simple of application for an industrial laboratory is that proposed by Berthier: it depends upon Welter's law, according to which the calorific power of different combustibles is proportional to the quantity of oxygen they combine with in burning; but this law is not exact: for instance, one part by weight of hydrogen requires, to change it into water, only three times more oxygen than one part of carbon requires to change it into carbonic anhydride, but the calorific power of the first body is four and a half times greater than that of the second. Berthier's method, therefore, will not give absolutely exact results, and the calorific power as thus given will always be found a little too feeble. The error is about ⅛th. Still the results obtained may be of use, for the question is not so much to determine the true calorific power of any one fuel as to compare the calorific powers of different fuels.

The direct determination of the quantity of oxygen necessary for combustion is difficult; we are, therefore, led to employ an indirect method. We heat a small quantity of the fuel to be tested with an easily reducible oxide, such as oxide of copper or more usually oxide of lead. By this treatment the solid fuel employed burns entirely at the expense of the oxygen of the metallic oxide with which it is mixed. By ascertaining the amount of metal reduced from

20

the oxide, we can calculate the quantity of oxygen which has been necessary for combustion of the fuel.

The oxide of lead is preferable to any other, on account of the facility of fusing the reduced lead or the oxide itself. When the assay is over, we can thus easily collect the metal at the bottom of the vessel in which the operation has been performed.

The experiment is made with one gramme of the fuel very finely pulverized, and mixed with 30 to 40 grammes of litharge, also in fine powder; the mixture is carefully placed in the bottom of an earthen crucible, and 20 to 30 grammes of pure litharge placed over it. A small porcelain mortar is best for mixing the materials; and it should be cleaned out afterwards, as well as the pestle, with some litharge, which should be used afterwards to cover the mixture in the crucible. The litharge is then covered with a layer of 4 to 5 millimetres of powdered glass, which must be free from lead or other easily reducible metals; the crucible, on a stand (*fromage*) and covered, is placed in a small blast furnace (Sefström's), unkindled coke heaped round, and covered with incandescent charcoal. The fire is thus kindled, commencing at the top; the mixture becomes heated in the same direction, so that there is no apprehension that any volatile part of the fuel to be assayed can escape the oxidizing action of the litharge. Heat for about an hour, progressively raising the temperature to a bright-yellow heat. The volatility of lead must be taken into considera-tion; to avoid the errors which that might occasion, we should not continue the heat after the reduction of lead is complete, nor raise it needlessly high, and make all our experiments under the same conditions as to duration and temperature. If the heat be too rapidly applied, the litharge may fuse before the coal is entirely oxidized; in this case particles of the fuel, owing to their smaller density, would raise to the surface of the melted mass, and would burn there without acting upon the litharge, even though very little air had access. The crucible in which the experiment is made should be at the most but half filled; without this precaution there would be considerable loss from the swelling of the mixture. When the mass is melted, withdraw the crucible from the fire, let it cool, then break it, take out the button of reduced lead and weigh it.

If the assay be well performed, the button detaches easily, has clean curved surfaces, and can be easily flattened under the hammer; if not so, the assay should be repeated. Two or three repetitions of experiments are indispensable, and they should not be considered good unless the weights of the buttons of lead do not differ more than one or two decigrammes from each other. Take the mean of the various results, and by this, calculate the value of the fuel; this value is expressed either in pure carbon, upon the basis that 34 parts of lead correspond to one part of carbon, or in units of heat (*calories*), it being granted that each part of lead corresponds to 230 French or to 912.65 British units.

V.—TESTING COAL WITH RELATION TO ITS PRODUCTION OF COKE.

It may be necessary to determine the amount and the quality of coke produced by a particular kind of coal; or the proportions in which different sorts of coal should be mixed in order to obtain coke of good quality. In either case, put 30 to 50 grammes of the coal or the mixture to be tested into a crucible; provide the crucible with a cover, having a small aperture for the escape of gases; lute the cover carefully to the crucible, and put it into the wind-furnace. Arrange the fire as for an iron assay. When the crucible or crucibles (for several experiments may be performed at once) are cooled, break them; weigh the coke obtained, and examine its quality. The difference between the weight of the coal employed and that of the coke corresponds to the amount of gas and volatile products expelled by heat.

The manner in which the assay has been heated, and the duration of the operation, influence the quality and quantity of the coke. If the cover be not well luted, the air may enter, and by burning some of the coke diminish its amount. Nevertheless the luting of the covers must preserve a small aperture for the escape of the volatile ingredients of the coal. It is necessary always to operate under like conditions.

Laboratory experiments do not furnish results perfectly according with those conducted on the large scale, but the hints they give are useful, especially when comparing different fuels.

Berthier's method is in reality little more than a useless chemical curiosity. It affords no real measure of the value of a fuel for industrial purposes. That value depends mainly in practice upon two conditions: What is the absolute calorific value of the fuel assumed, all perfectly burnt, and without waste? What are its physical and mechanical properties upon which its advantageous use for special heating purposes depends? Thus, if for raising steam, etc., its volume for unit of weight, its caking or not, its coherence while consuming, are important elements: if for smelting operations, these also apply. But in addition, is the important question, what degree of hardness the coal or the coke from it possesses; will it in the blast furnace, for example, bear a certain amount of " burden ?"

These latter properties must chiefly be judged of by tact and experience, prior to actual trial on the great scale. By suitable arrangements and the employment of various precautions, it is possible to determine by means of the mercurial calorimeter (of Favre and Silbermann), with minute accuracy, the absolute calorific value of any sort of fuel, solid or liquid; but the operations are so delicate and tedious as to be wholly unsuited to industrial use.

But by two experiments we can calculate the absolute calorific value of any coal fuel without much difficulty. By one we determine the percentage of incombustible ashes; by another, conducted exactly as an organic analysis, or as described for the combustion of carbon liberated from cast-iron, we determine the elementary composition of the dry fuel—*i. e.*, the percentage of oxygen, hydrogen, and carbon (neglecting the sulphur, which in nearly all industrially valuable fuels is present in very minute proportion). Favre and Silbermann have determined the following calorific values for one pound (avoir.) of hydrogen and of carbon :—

	lbs. of oxygen to burn.	lbs. of atmospheric air to burn.	Total heat in British units.
Hydrogen	8	36	62,032
Carbon burnt to carbonic oxide	$1\frac{1}{3}$	6	4,400
Carbon burnt to carbonic acid	$2\frac{2}{3}$	12	14,500

Now it has been also proved, that the total heat of combustion of any compound of carbon and hydrogen, is the sum of the quantities of heat, which each of these elements if burnt separately would produce. But it is also to be observed that where oxygen is also present in the compound (as it is more or less in nearly all fuels), so much of the hydrogen as is required to form water with the contained oxygen (viz., eight parts of oxygen for one of hydrogen) is burnt without producing any final calorific effect. Therefore, deducting from the pound of fuel the weight of the incombustible matter, nitrogen (if present) and ash, and expressing by C, H, and O the three elements present in decimals of a pound, we have the following general formula, by which, after having made the two experiments above described, we can calculate the calorific or total heating power (h) of any fuel:—

$$h = 14{,}500 \left(C + 4.28\,H - \frac{O}{8} \right) \quad . \quad . \quad . \quad \text{I.}$$

Or the theoretic evaporative power (E) of one pound of the fuel in pounds of water boiled off at one atmosphere from 212° Fah., is:—

$$E = \frac{h}{96} = 15\left(C + 4.28\,H - \frac{O}{8} \right) \quad . \quad . \quad . \quad \text{II.}$$

Of these theoretic values, however, much is lost by the necessity of passing in the atmospheric air much nitrogen (which must be uselessly heated) through the fuel, besides a larger volume than is necessary to afford the oxygen barely necessary to combustion, and by other conditions of loss of heat applying more or less to all furnaces.

As regards assays for coke, it is quite impracticable to judge with any certainty what may be the industrial or commercial value of a given coal for coking purposes, except by trials conducted on the great scale, and in the actual coke oven.

We can merely decide by assay on the small scale, whether the coal shall coke at all, or if so, give a worthless coke, or one worth more or less. To make this trial a luted crucible is a bad apparatus, and the wind furnace, an uncertain and unequal apparatus for heating it.

A small wrought-iron or cast-iron retort, holding one or two cubic feet of the fuel, and embedded in an exactly regulatable furnace of its own, is the best procedure ; and with such apparatus very tolerable indications as to probable quality of coke when made on the great scale, are possible. Still, coke from the *same coal* is generally better and harder as the scale of oven in which it is made is greater.

[The rapidity with which bituminous coal is heated, and also the intensity of the temperature, have a great influence in the manufacture of coke. Certain kinds of coal, poor in bitumen, require to be suddenly seized by an intense heat in order to retain the bitumen which agglutinates them. Such coals, if heated slowly at a moderate temperature, would not coke, and would remain quite pulverulent.

On the other hand, highly bituminous coals may be coked slowly at a comparatively low temperature.

The hardness, density, and capability of supporting a "burden" without crumbling to pieces, depend not only on the temperature of the coking operation, but also on the thickness of the layer of coal; the weight above, compressing the coking material and increasing its density, at that period during which it is in a pasty state. The coke from gas retorts, although of the most bituminous kind of coal, is porous and wanting in hardness, because it is spread in thin layers in the retort, and there is no weight above it to compress it when it swells.

In our opinion, an apparatus for trying the coking properties of bituminous coals on a small scale, should be arranged so as to be heated rapidly or slowly, as desired, and to have the coal in a somewhat thick layer. A metallic vertical retort may answer the purpose.]

SUPPLEMENTARY NOTES BY THE AUTHORS.

———◆———

I.—ASSAY OF SUBSTANCES CONTAINING ZINC.

THE volumetric method being the only one employed for the industrial assay of these substances, we shall not describe any other. The plan of our work will not permit us to dilate upon processes of complete analysis, generally very tedious; for the composition of zinc ores, *cadmie* (*i. e.*, flue-dust), etc., is almost always more complex than that of iron ores.

Zinc ores, and those other substances containing zinc which we may have to assay, contain this metal in the state of an oxide, sulphide, carbonate, or silicate. They may contain variable proportions of oxides of iron, manganese, lead, copper, and cadmium, the sulphides and carbonates of these metals, various silicates and gangues insoluble in acids, and, finally, carbonates of calcium and magnesium, and sulphides of antimony and arsenic.

If the substance to be tested does not contain a notable quantity of lead, copper, cadmium, or manganese, the assay is made in the following manner: Treat 0.5 gr. to 1 gramme of the substance with hot hydrochloric acid, to which add, towards the end of the operation, some drops of nitric acid, in order to change the protosalt of iron which may exist in the solution into a persalt; then evaporate to dryness, and treat again with a few drops of hydrochloric acid and a small quantity of hot water. Having neutralized with ammonia, add an excess of carbonate of ammonium in solution. Heat to collect the precipitate; let it deposit, then filter and wash with hot ammonia water. Almost all the zinc is kept in solution by the excess of ammonia employed; nevertheless, the precipitate containing the peroxide of iron and alumina still retains a small quantity. In order to obtain the whole of the zinc, dissolve the precipitate with hydrochloric acid, and treat the solution in the same way as the first. Wash the precipitate again with ammonia water, and continue the washing until a small portion of the filtered liquid tested with hydrosulphide of ammonium remains perfectly clear. The small quantity of zinc lost from the mass for the purpose of this testing has no perceptible influence on the final result. The two ammoniacal liquids are then united.

If the ore contain lead, add sulphuric acid to the first solution before evaporating to dryness, and treat the residue with dilute sulphuric acid. The lead remains as a sulphate with the silica and lime; let it deposit, filter, wash, and then treat the liquid with carbonate of ammonium.

We may also abstract the lead by treating the acid solution with a current of hydrosulphuric acid. This process is necessary when the ore contains copper. Lead and copper are then precipitated as sulphides;

separate the precipitate by filtration, evaporate the filtered liquid to dryness, taking care to add, towards the end of the operation, a small quantity of nitric acid, in order to peroxidize the iron which the hydrosulphuric acid had deprived of its oxygen. The rest of the operation is the same as above described.

A simple method of separating the lead and manganese at the same time is to add to the ammoniacal solution, before separating the precipitate of oxide of iron, etc., a few drops of a solution of phosphate of sodium. Insoluble phosphates of lead and manganese are formed; the zinc remaining in solution.

When the ammoniacal solution is prepared, we proceed with the volumetric determination of the zinc it contains. For this purpose several methods have been proposed : one only has been successfully practised—that based on the faculty that oxide of zinc in an alkaline solution possesses of forming a white precipitate of sulphide of zinc under the action of sulphide of sodium. By dropping from a graduated burette a standardized solution of this reagent into an ammoniacal liquid holding zinc in solution, we learn the quantity of the latter from the volume of the solution of the sulphide required to effect complete precipitation. The precipitate of sulphide of zinc does not deposit rapidly enough to allow us to judge of the end of the reaction by the cessation of the precipitation; it is necessary to look for another indicator. Several have been proposed. We shall mention four, which are all based on a change of color produced by the solution of zinc when it contains an excess of sulphide of sodium.

1st. *By Hydrate of Peroxide of Iron.*

Add to the ammoniacal solution of zinc two or three drops of a solution of perchloride of iron; red-brown flakes of hydrate of peroxide of iron, which fall to the bottom of the vessel, are formed. Under the action of sulphide of sodium these flakes become gradually darker colored, greenish, and finally black; the sulphide of zinc does not produce this reaction on the hydrate of peroxide of iron, but the latter once blackened, does not again become red under the action of an ammoniacal solution of zinc: we must take care that it does not come in contact with the sulphide of sodium *before* all the zinc be precipitated. We succeed in doing this by letting the flakes of hydrate of peroxide of iron settle to the bottom of the vessel before each new addition of the standard solution. We may consider the reaction ended the moment the hydrate of peroxide of iron becomes gray or black; but all these assays should be performed in the same way and in the same space of time. It has been proposed to replace the hydrate of peroxide of iron, which, on account of its lightness, rises easily, by a strip of paper soaked with the perchloride of iron, and ballasted with a glass rod or a platinum wire; or by slips of unglazed and half-baked white porcelain soaked in solution of the same salt.

2d. *By Chloride of Nickel.*

When from the appearance of the precipitate of sulphide of zinc, or by the quantity of the standard solution of sulphide of sodium employed, we judge that the reaction is nearly at an end, we should then add the reagent by only a cubic centimetre at once and finally by tenths of a cubic centimetre at a time, taking care on each occasion to shake the liquid well. After each addition let fall a drop of the liquid on a white porcelain plate, and to that add one drop of a diluted solution of chloride of nickel. As long as there is no excess of the sulphide, the extreme edge of the drop of chloride of nickel remains blue or green; but with the least excess it becomes grayish-black, and even deep black if the sulphide be in notable excess.

3d. By the Nitroprusside of Sodium.

Operate exactly as with the chloride of nickel. The sulphide of sodium possesses the property of producing with the alkaline nitroprussides a rose, violet, or red-purple color, according to the degree of concentration of the liquids. The reaction has terminated, when the addition of one drop of the nitroprusside to one drop of the liquid assayed, produces one of the colors just indicated.

4th. By Paper prepared with a Salt of Lead.

Salts of lead having the property of becoming black under the action of sulphide of sodium, we may make use of them to recognize the presence of this combination in the solution of zinc. As we cannot employ a solution of a salt of lead, because the sulphide of zinc colors it, we use paper impregnated with or covered with a salt of lead as described under Test Papers. To ascertain the end of the reaction, wet the paper with a drop of the assayed solution. If it contain sulphide of sodium, it will produce a brownish-yellow spot, or even a black one if the sulphide be in excess.

In order to ascertain a minute excess of sulphide; let a larger drop fall on the first drop; the least difference of shade indicates the end of the reaction.

Fixing the Standardized Solution.

The solution of sulphide of sodium is made by dissolving 25 grammes of the crystallized sulphide in a quantity of water sufficient to obtain one litre of liquid. We determine the standard by means of a solution of 2 to 5 decigrammes of pure zinc in a small quantity of dilute hydrochloric acid. Having added to the liquid an excess of ammonia and carbonate of ammonium, pour in the solution of sulphide of sodium until the zinc be completely precipitated, which we can ascertain by one of the methods above given. It is well to operate with the quantities of zinc and the combinations of ammonium, and a volume of liquid similar to those we employ in assaying ores, that we may be as nearly as possible in the same conditions for each operation.

The end of the reaction being indicated by a slight excess of the sulphide of sodium, it is necessary to ascertain this excess by a special test, which is made in the following manner: Drop the sulphide into a

21

volume of aqueous solution of ammonia equal to that used in the assay, until the indicator marks the end of the reaction with the same intensity as at the time of the determination of the standard. The excess of sulphide thus ascertained must be subtracted every time from the volume of the standard solution employed.

II.—ASSAY OF SUBSTANCES CONTAINING LEAD.

The assay by the dry method is the only one made use of in industrial laboratories, and sales are made only in accordance with the results it furnishes, though these results do not express the real proportion present in the substance assayed.

Alkaline fluxes are always used, as they alone produce easily fusible slags, and permit rapid operations at a relatively low temperature: without observing this condition, we should be liable to obtain erroneous results, owing to the volatility of the lead. We cannot detail here all the mixtures which may be used, nor all the modes of conducting the assay, as such would exceed the limits of this work; we shall merely describe the method most commonly practised in industrial laboratories for analyzing these ores.

The assay is made with 10 to 30 grammes of the substance finely pulverized. If the ore contain sulphur, mix it with once or twice its weight of dry carbonate of sodium, and put the mixture into a paper cornet or case. In a wind-furnace, which should not be very deep, heat a crucible of forged iron to redness, and clean its interior with fused carbonate of sodium; heat it again, and when it has attained the proper temperature, a yellow heat in daylight, put in the paper cornet containing the mixture to be acted upon. Cover the crucible, and heat it for about five minutes without closing the furnace, then urge the fire for ten minutes, so as to obtain at the end a temperature sufficiently high to render the mass very fluid, and in a state of tranquil fusion. Then withdraw the crucible from the fire; cause the grains of lead possibly adhering to the sides, to fall to the bottom, and unite with the button, by shaking the crucible and striking it lightly upon the floor; then pour the whole into an iron ingot mould in the shape of a cone, with the apex downwards. We may also pour first the greater portion of the scoria on to a metal plate, and only make use of the ingot mould for the last part containing the lead. As soon as the slag has solidified, a blow of a hammer sufficies to separate the button of metal; clean the latter by brushing it, and subjecting it for a few minutes to the action of dilute sulphuric acid, then dry and weigh it.

If the ore contain earthy matter, gangues, or oxide of iron, add to the carbonate of soda 4 to 5 grammes of powdered glass of borax: this reagent serves to vitrify the lime, magnesia, etc., which, without that, would remain suspended in the scoria, and diminish its fluidity. When oxidized ores have to be assayed, the flux should be a reducing one; employ for this purpose either black flux, or a mixture of carbonate of sodium and crude tartar (*Argol*) in the proportion of two parts of the first to one of the second. Add borax to this mixture, if the ore contain matter not vitrified by the alkaline carbonate alone. The assay is performed in the same manner as that for sulphuretted ores, with carbonate of sodium.

Reducing fluxes may also be used with sulphurous ores. If the ore contain silver, it passes into the button of reduced lead, whence, if we want to determine its amount, it must be extracted by cupellation, or by solution of a known weight of the lead button in nitric acid, and precipitation of the silver from the largely diluted solution by hydrochloric acid. Cupellation, when thoroughly performed, is, no doubt, the most exact method of fixing the standard of silver contained in lead. Directions have been minutely given by Plattner and others for its determination by blowpipe assay, by cupellation, but they can scarcely be viewed with perfect confidence.

The range of affinities of iron is so great, and its tendency in the metallic state to combine with and hold tenaciously, small amounts of many foreign elements is, as already remarked, such that either in the state of its ores or in those of its three great industrial metallic conditions—cast-iron, wrought iron, and steel—to which we may add meteoric iron (which may hereafter prove to have also a terrestrial existence), that it is found in one state or another occasionally in combination with a tolerably large proportion of all the sixty-four known elementary bodies.

A complete treatise upon the analysis of iron and its ores, etc., is therefore nearly co-ordinate with one upon analysis in general. Such, however, is far beyond the scope of this little work, which is not designed for pure scientific research, but for the humble use of the industrial laboratory.

In its ores we encounter iron, constantly in combination with silica, alumina, lime, magnesia, and manganese; more rarely with carbon (or carbonic acid), sulphur, and arsenic; and, if we include meteorites, in combination with nickel, cobalt, and chromium.   We also find it widely diffused in combination (occasionally with zinc) with titanium as well as with chromium. But the first seven bodies named are those only with which the sidero-technist is concerned; and these are all dealt with in the preceding pages, besides some of the others.   In the metallic state we nearly always find iron associated with carbon (combined and as graphite), silicon, sulphur, phosphorus, and manganese. These are the elements of most industrial importance.   Occasionally we find it combined with aluminium, calcium, magnesium, arsenic; more rarely with tungsten, titanium, vanadium, and (when tungsten is present *from the ore*) probably tin is never quite absent.   Nitrogen has been presumed present in some steels, and probably all iron occludes certain gases in solidifying from fusion, as shown by Graham.   More rarely still traces of chromium, nickel, or cobalt have been observed in certain makes of pig-iron, and even lead ; whilst in iron resulting from ores, produced by the gradual natural decomposition of pyrites, or containing pyrites, it is all but certain that gold or silver, or both, may exist in extremely minute quantity.   But with the greater portion of these bodies the industrial chemist or assayer is not concerned.   Still in the industrial laboratory, he may be at any time called upon to determine qualitatively or quantitatively any one or more of these numerous elements, and if wholly unaccustomed to their reactions, must find himself much perplexed.   For the study of mineral analysis generally, and in its largest special relations to iron (in addition to the standard works on General Analysis of Rose, Rammelsberg, Frésénius, etc.), no industrial laboratory should be without the great work, "Docimasie," etc., of M. Rivot, already referred to.

The determination of aluminium and titanium in cast-iron will be found well described by Rivot, t. iii. p. 538 ; of Vanadium,

p. 540; and of tungsten (now entering industrially into the manufacture of steel) at p. 547 of same volume. These bodies thus referred to have been but cursorily treated of in the text, and properly so, by reason of their rare occurrence.

Nor does this work at all treat of the analysis of furnace or other gases, which, however, must sometimes devolve on the ironwork chemist; for this, amongst other works, Bunsen's "Gasometrie" should be studied.

APPENDIX.

BY THE AMERICAN EDITOR.

IRON ORES.

WE call *iron ores* the natural compounds of iron, which are employed in the manufacture of wrought and cast-iron, and steel. Many minerals holding iron in some form of combination, will be passed over, or examined but slightly, because they are simply minerals more interesting to the mineralogist and geologist than to the iron manufacturer.

Iron ores are classified either geologically or chemically. For instance, the primary or crystalline ores will comprise the magnetic and specular red oxides of the primary rocks. There will also be the crystallized carbonates and brown hematites of the secondary formations, the fossil ores of the upper Silurian rocks, the carbonates of the coal measures, and the bog ores in sands or marls of tertiary deposits. This arrangement presents the advantage of leaving together different kinds of ores which are more or less mixed one with the other, of facilitating geological research, and of giving a certain amount of knowledge as to their purity, since, as a general rule (with some exceptions of course), the older and the more crystallized an iron ore is, the purer it is expected to be.

On the other hand, the chemist in his laboratory, and

the metallurgist in his works, prefer arranging the ores according to their chemical composition, and in this manual we shall adopt the chemical classification, remembering, however, that a mineralogical and geological knowledge of ores, is often useful to the chemist.

The *iron ores* are oxides, presenting various degrees of oxidization. The protoxides are combined with carbonic anhydride, and form the CARBONATES. The combination of one atom of protoxide ,with one of peroxide, results in MAGNETIC ORES. Lastly, the peroxides, whether anhydrous or hydrated, form the RED HEMATITES and the BROWN HEMATITES.

These subdivisions are useful, although it will be seen further on, that many carbonates and magnetites are more or less peroxidized, and that many red hematites contain more or less of brown oxide, etc.; these changes being due to the never-ceasing action of the atmosphere, of water, carbonic and ulmic acids, etc. We are obliged to give typical examples of ores, which in practice will be found more or less altered in appearance, streak, and composition; nevertheless, with judgment and a few tests, the chemist will be able to ascertain the classes to which the ores which he has under examination belong.

It cannot be expected that this appendix will comprise a complete vocabulary of all the names given to iron ores, and which often vary with each locality, with appearance, color, etc. We mention the geological formations in which the iron ores are *generally* found, only so far as may be useful to the chemist in his researches, and for more extended information on the subject, we refer the reader to *Lesley's Iron Manufacturer's Guide*, *Dana's Mineralogy*, the various Reports of State Geological Surveys, etc. etc.

CARBONATES OF IRON.

Siderite, chalybite, spathic, and *sparry iron, steel ore.*
The crystals belong to the rhombohedral system, with the
faces often curved. Very frequently in crystalline masses,
which are sometimes fibrous, lamellar, or scaly. Lustre,
vitreous, and more or less pearly. Color, ash-gray to
brownish-red, according to the amount of peroxidation.
Streak, white or gray, or more or less brown, from the
presence of a greater or less quantity of peroxide. The
color of the streak is ascertained either by scratching the
ore with a pointed tool or a file, or by rubbing it upon a
piece of unglazed porcelain. Sp. gr. 3.7 to 3.9. Composition, carbonic anhydride 37.9 + protoxide of iron 62.1 =
100, which is equal to 48.22 per cent. of metallic iron. It
is slowly soluble in dilute hydrochloric acid, and decrepi-
tates and becomes magnetic by heating.

Spherosiderite is the same ore in globular concretions.

It happens very often, that part of the protoxide of
iron is substituted by a greater or less proportion of
protoxide of manganese, magnesia, or lime.

For instance, *oligonite* contains 25 per cent. of protoxide of manganese. The streak is yellowish-white, and the
mineral is phosphorescent when heated.

Sideroplesite is formed of two atoms of carbonate of
protoxide of iron, and one of carbonate of magnesium.

Siderodot contains a large proportion of carbonate of
calcium.

The crystallized carbonates occur generally in gneiss,
mica slates, etc., and are often associated with other ores,
such as the sulphides of copper, iron, lead, silver, etc.
The phosphates are seldom met with, and the sulphur,
quartz, and copper are the greatest impurities the iron

master has to contend with. With care in picking up and roasting these ores, they are excellent for the manufacture of steel, and often contain a large percentage of manganese.

The crystalline varieties are found in England, in Durham, Cornwall, North Devon, and Somerset; and in Germany, in the country of Seigen. The accompanying rocks belong to the Devonian formation.

In Styria, they are found in micaceous and talcose schists, and metamorphic limestones.

The highest parts of the lodes are generally changed into brown hematites, and the lamellar or crystallized form often remains unchanged. Examples have been found of further transformation, by which the brown oxide becomes red hematite, then specular iron, and lastly magnetite, by deoxidation due to organic substances.

In the United States, the crystalline carbonates (although found in many parts of the country) are not employed to a very great extent. Up to the present time, the deposits have not been found of sufficient magnitude to be the basis of a special iron manufacture.

The crystallized carbonates in layers found in the south of France (Gard), mark the transition from the crystallized to the lithoid form. They are nearly black, and mixed with quartz and clay. They are intersected by veins of siderite and of sulphate of barium, and contain metallic sulphides, but scarcely any of the carbonates of manganese, calcium, or magnesium.

The *earthy, lithoid, argillaceous,* and *calcareous carbonates of iron* are very abundant in England, and quite so on the Continent, and in the United States.

The *oolitic carbonates* resemble oolitic limestone in structure.

An *argillaceous carbonate*, which is white, and mixed

with clay, is found in Maryland, and extends for fifty miles west of Chesapeake Bay.

The limestones and clay slates of coal formations present numerous varieties of earthy carbonates.

The *calcareous carbonates*, *limestone* or *buhrstone ores*, contain a large proportion of carbonate of calcium mixed with that of iron. Their color is very variable, gray, grayish-blue, brown, and bluish-black. They are found in Pennsylvania, Ohio, Indiana, Maryland, and Kentucky.

Clay ironstone (which by the by is a common name given to many different kinds of iron-ores) is a carbonate of iron found in the clay slates, sandstone, shales, and all kinds of sediments of the coal measures, and even in the coal itself, mixed with a great deal of slate. It forms irregular, and flat strata, which contain many impurities derived from the surrounding rocks. The phosphates of calcium and aluminium are nearly always present, and impair the quality of the metal considerably. Arsenic and sulphate of calcium are also found in these ores.

The *nodular carbonates* of the same formations are compact and intimately mixed with clay and a certain proportion of the carbonates of calcium, magnesium, and manganese. When concretionary, they often contain fossils of fishes, shells, and plants. The fresh fracture is yellow, gray, or bluish, but becomes brown by exposure. They are less impure than the lithoid clay iron-stones, although they still contain some phosphorus, and from 0.35 to 2.40 per cent. of organic matters.

The *black band ironstone* of England is of a dark brown or black color, and often presents a shaly structure. After calcination, effected in heaps and without additional fuel, it yields from 40 to 60 per cent. of metallic iron. In its crude state, it contains from 40 to 47 per cent. of protoxide of iron, with more or less of peroxide, 1 to 3 per cent. of oxide of manganese, 7 to 11 per cent. of organic matter, and but a small proportion of phosphorus. It produces a good metal.

The *Cleveland ironstone* of England is of a dull or dark-bluish-green (due to some silicate of iron), and contains fossil shells of the genera *pecten* and *avicula*. It is oolitic in structure, and is found in the middle Lias. The carbonate of iron is more or less decomposed, some parts being even magnetic, and resembling the *chamoisite* found in Switzerland. It contains little, if any, organic matter, and less sulphur and manganese than the black band ore. On the other hand, the proportion of phosphorus is greater.

Of the various kinds of ironstones of the coal measures, many are found and used in the United States. However, up to the present time, the real black band, rich in metal, good in quality, and having sufficient bituminous matter to be calcined without additional fuel, still remains an ore peculiar to the British Islands. It is said that two-thirds of the iron production of England comes from these iron stones of the coal measures, which are simultaneously worked with the coal seams.

MAGNETIC ORES.

Magnetite, oxidulated iron in octahedral and dodecahedral crystals, with a sp. gr. varying between 4.9 and 5.2. Color black, with often a metallic lustre, coarsely and finely crystalline, and massive, magnetic, and generally magnetipolar to a greater or lesser degree. Streak black when pure. Composition, oxygen 27.6 + iron 72.4 = 100; or protoxide of iron 31.03 + peroxide 68.97 = 100. Part of the iron is often substituted by a certain proportion of magnesia or titanic acid, which diminishes the magnetic properties of the ore.

Instead of the normal ratio of 1 : 1 between the protoxide and peroxide (ferrous and ferric oxides) of iron, many crystalline samples contain a greater proportion of peroxide. This is also found in the massive varieties, and the streak, instead of being pure black, has a reddish or brownish cast about it.

Ochreous magnetite is black and earthy, and part of the protoxide of iron is replaced by the protoxide of manganese. A little copper is also found in it.

Magnetic sands are found on the coast of Labrador, in Canada, Long Island, New Zealand, in the residues of gold-ore washings, and are the product of the erosion of rocks. Their composition is very variable, and comprises specular and magnetic ores, titaniferous and chrome irons, garnets, zircons, silicates, etc. Their yield in metallic iron is sometimes considerable.

Magnetic iron-ores occur mostly in crystalline metamorphic formations. The gangues are serpentine, various kinds of pyroxenic rocks, chloritic and micaceous slates, garnets, quartz, feldspar, corundum, etc. The texture of these ores varies with the containing rocks ; the most massive are found in talcose schists, and the most crystalline in hornblendic gneiss and crystalline limestone.

As a general rule, with some few exceptions, magnetic ores are excellent for the manufacture of iron. They seldom contain arsenic, and very little, if any, phosphate intimately mixed. Apatite, or crystallized phosphate of calcium, sometimes accompanies certain coarsely crystalline varieties of magnetite, and may be perceived with the eyes. The sulphides of iron and copper are more frequently met with in certain localities.

Magnetites are often decomposed into brown hema-

22

tites on the surface, and also contain specular iron in the state of intimate mixture, as is seen in certain Marquette ores of Lake Superior.

Magnetic iron-ores are most abundant in Siberia and Elba; at Arendal, in Norway; at Fahlun, Damemora, Taberg, etc., in Sweden; and in the United States, in the Adirondacks, in Pennsylvania, New Jersey, Maryland, California, in North Carolina, and other Southern States, Missouri, on Lake Superior, etc. etc.

The geological disposition of the ores in Sweden tallies remarkably well with that of the same ores in the Adirondacks and North Carolina, that is: in gneiss alone or accompanied by granite, and talcose, chloritic, and micaceous slates, and in hornblendic rocks intercalated in the gneiss.

Red Hematites.

These iron ores are also known under the general names of *anhydrous peroxides* and *red oxides*.

Specular iron, oligist iron, iron glance. Rhombohedrons and derived forms, the primitive one being rare. Also lamellar and tabular. Lustre metallic, and sometimes splendent and iridescent. Specific gravity 4.5 to 5.3. Color from a dark-steel gray to iron-black. Streak gray, more or less violet, and sometimes with a brown tinge when the red oxide is mixed with a certain proportion of hydrated peroxide. It is occasionally slightly magnetic, from the presence of magnetite. Some samples contain titanic acid and magnesia. Composition, oxygen 30 + iron 70 = 100.

Micaceous iron has a foliated or micaceous structure, with a steel-gray metallic lustre, when the scales are large enough. A variety, with very small scales, resembles graphite greatly, although, when finely ground, its powder is red. It is found in quartz rocks and micaceous schists, not in large deposits, and generally contains manganese and a large proportion of quartz.

Martite crystallizes in octahedrons like magnetite. Its color is iron-black, and its streak is red, with more or less of brown and purple. It is sometimes imbedded in massive hematites.

Violet ore of Belgium is intermediary between the crystalline and compact varieties of red oxide. It forms large beds, and has a schistose texture. Color violet-red to iron-gray. It is intimately mixed with quartz and clay, and seldom contains limestone or sulphate of barium. Manganese and phosphate of iron are found occasionally. It is a valuable ore, and produces a good metal.

Compact, columnar, and *fibrous red hematites* are so named according to their structure. Color brownish-red to iron-black. The fibres are often silky and radiating. A compact variety, *bloodstone*, is so dense and hard that burnishing tools are made with it.

Iron minium, red ochre, or *red chalk*, is an earthy peroxide, mixed with clay.

Specular schist, or *itabyrite*, is a schist containing a large quantity of micaceous iron, and resembling mica schist. Found in North and South Carolina.

Clay iron stone, argillaceous hematite. Color brown-red to brown-black. Intimately mixed with sand and clay, and very hard and tough. A red variety, with jasper-like structure, is known under the name of *jaspery clay iron ore.* Another variety, the *specular schist* or *slate iron*, found near Marquette, Michigan, in Huronian rocks, is quite siliceous, and contains small crystals of martite. It bears great resemblance, in schistose structure, to the Pilot Knob ores of Missouri.

Reniform (kidney) ores exist in Bohemia, in the Hartz Mountains, and in Cumberland, England, in the shape of hard, botryoidal masses, without metallic lustre.

The red hematites, of Cumberland, Ulverstone, and Lancashire, England, employed for the manufacture of Bessemer pig-metal, exist in irregular deposits in the carboniferous limestones. They yield traces of sulphur and phosphorus, but are rather siliceous.

Puddler's ore, from Cumberland (England), is a compact and unctuous micaceous variety, and is used for the lining of puddling furnaces.

Lenticular iron ore, oolitic fossil ore, fish egg ore. Oolitic in structure, but in certain places the grains are larger and flattened. An argillaceous variety is found in New York (Oneida), Pennsylvania, Alabama, Virginia, Georgia, Tennessee, and Kentucky, in the upper silurian formations. These fossil ores are more or less impure; they contain limestone and sometimes a large proportion of silica and alumina. There are traces of manganese.

In Europe oolitic and fossiliferous beds of red oxide, exist in Champagne and Berry, in Jurassic, and the middle tertiary rocks. Near Cardiff, Wales, and in Belgium, at the base of the carboniferous limestone.

The crystallized and compact red hematites occur in the same metamorphic rocks, especially mica schists, in which magnetites are found. Indeed, these two iron ores are sometimes mixed together, and are frequently accompanied by a certain proportion of brown hematite. Chromic iron, sulphide of nickel, and other metallic sulphides, and limestone, pure or magnesian, are seldom found in the older ores, which sometimes contain manganese and sulphate of barium.

These ores are widely diffused, and are found in Africa, Elba, France, Norway (Arendal), Sweden, Germany,

England (Cumberland, Lancashire, etc.), Brazil, Chili, the Adirondacks, and elsewhere in New York, Missouri, Lake Superior, New Jersey, New Hampshire, several of the Southern States, etc. etc.

BROWN HEMATITES.

The hydrated peroxides are the most widely diffused iron ores. They are found in nearly all kinds of formations, and their form and qualities are also very variable. The color of their streak is from a reddish to a yellowish-brown, but brown predominates. They all contain combination water, and in some of them this water exists in chemical proportion.

Turgite. Fibrous, with fibres possessing a satin-like lustre; sometimes botryoidal, stalactitic, or dull earthy. Color red to reddish-black. Streak brown-red. Specific gravity 3.56 to 4.49. Composition, sesqui-oxide of iron 94.7 + water 5.3 = 100. It often contains manganese, and seldom copper, lead, and sulphuric acid. It flies to pieces when heated, and often associates with ordinary limonite, being harder than this latter ore, and being separated from it by quite a distinct line of demarcation.

Göthite, pyrrhosiderite, brown iron stone or *ore.* Prismatic, with longitudinal striæ; also fibrous, massive, reniform, and stalactitic. Specific gravity 4.0 to 4.4. Not much lustre. Color yellowish-red to blackish-brown. Streak brown, more or less yellow. Composition, sesqui-oxide of iron 89.9 + water 10.1 = 100. Contains manganese often, and copper and lead seldom. Other varieties are named *needle iron stone, lepidocrocite,* etc.

Limonite, brown ochre. Stalactitic or botryoidal with a fibrous structure; also massive, concretionary, and earthy. The lustre is silky, for massive varieties, and also sub-metallic or dull earthy. Specific gravity 3.6 to

4.0. The fracture is dark-brown, the exterior surface often black, and the streak brown with a more or less yellow tinge. Composition, peroxide of iron 85.6 + water 14.4 = 100. Some of the varieties of limonite are *bog ore*, loose, porous, holding often petrified leaves, etc.; *brown-clay iron stone*, compact and often nodular ; *pisolitic ore*, like an aggregation of small peas; *oolitic ore*, like an aggregation of fish eggs ; some kinds of *ochres*. The purity of these ores is very variable.

Xanthosiderite, in needles, stellar, and concentric, also in the ochreous form. Color, yellowish and reddish-brown. Composition, sesquioxide of iron 81.6 + water 18.4 = 100. Often contains manganese.

Limnite, similar in appearance to the previous ore, has for composition: peroxide of iron 74.8 + water 25.2 = 100.

Ochres, brown or yellow. An intimate mixture of hydrated oxide of iron with more or less clay, fine sand, etc. The depth of their color should not always be taken as an index of their yield in metal, since some are rich enough to be melted, although they may appear quite lean (poor) to the eye.

Bean ore. A loose concretionary brown hematite filling cracks in limestone. Size, from that of a pea to a walnut, with a ferruginous sand or clay as cement.

Lake ore. Found in Sweden, Norway, and Finland. In granular concretionary forms, produced by infusoria (diatomaceæ), which separate iron from water. A production of our times. These ores contain phosphorus, manganese, organic matter, and but a small proportion of limestone.

Of these brown iron ores it may be said, in general, that the more recent their formation, the less pure they are. They result from the alteration of older ores and sulphurets under the influence of the atmosphere, water, carbonic anhydride, ulmic acid, etc. In the older rocks, the magnetites, carbonates, and red oxides are often covered with a cap of brown hematite. The brown ores of the secondary formations are found in the lias, in the oolitic and lower green sand rocks, impure and earthy. They sometimes result from the decomposition of earthy carbonates. In Belgium, in the valley of the Meuse, they are often mixed with lead and calamine, resulting from the surface decomposition of sulphides of iron, lead, and zinc, which, lower down, are found unaltered.

In some bog ores produced by erosion, all kinds of iron ores are found in admixture.

We shall now give *the French method of classifying brown hematites*, as it furnishes useful indications as to the impurities generally to be found in them.

The "brown hematites of old rocks" are compact, fibrous, and of a very dark brown. Often intimately mixed with a hydrated oxide of manganese, and with more or less quartz and clay. Limestone, pure or magnesian (dolomitic), is sometimes met with; but the sulphates of barium, the metallic sulphides, and the phosphates and arseniates are rare.

The "compact kinds of brown ores of more recent formations" lie often at the separation of two different kinds of rocks, or in Jurassic limestones. They are not so dark as the previous ores, and their fracture is irregular, and sometimes slightly fibrous. They contain little manganese, but more of limestone pure or magnesian, quartz, clay, and hydrated alumina. Sulphate of barium, and compounds of phosphorus, sulphur, and arsenic are seldom found in them.

The "earthy and geodeic or septarian kinds" result from the alteration of metallic sulphides. They form caps or hats on the surface, and often affect the shape of balls preserving the fibrous structure of the originary pyrite. These ores are sometimes very impure, and contain arsenic, sulphur, lead, zinc, phosphorus, sulphate of calcium, silica, limestone (seldom magnesian), basic sulphates of iron, etc.

The "kidney granular ores of tertiary rocks" are mixed with clay marls, which may be separated by washing. They do not contain much phosphorus, arsenic, sulphur, or manganese.

The " oolitic ores of oolitic limestones" are sometimes of a dull-blue color. They contain hydrated alumina, silica, marls, and silicates, also phosphates of iron, aluminium, and calcium. Some of the grains are magnetic.

The " recent bog ores" have sometimes a resinous fracture, no fibres. They contain silica, phosphorus, sulphur, arsenic, basic sulphates of iron, clay, and limestone. Manganese and sulphate of barium are seldom met with.

The " Ochres" are rich in clay and sand, and contain occasionally some sulphur, phosphorus, and limestone,

Miscellaneous.

Under this heading we place certain iron combinations, which could not be included within the preceding ores.

Metallic iron, meteoric iron. It has been said that very thin layers of metallic iron have been found in the mica schists of Connecticut, but this is more than doubtful and requires confirmation. On the other hand, meteoric iron has been found in many localities, in masses of all sizes, up to 15 tons weight. The surface is oxidized, but the inside possesses the ordinary color of iron and is crystalline in texture. All meteoric irons are alloyed with nickel, in the proportion of from 1 to 20 per cent. The other metals occasionally combined with it, are cobalt, chromium, manganese, arsenic, sulphur, and copper. A phosphide of nickel occurs also in many samples. After iron, nickel and cobalt predominate, and the alloy is malleable. Meteoric iron cannot be called a commercial ore, but it is difficult to pass in review the principal compounds of iron without mentioning it.

Franklinite. In dodecahedral and octahedral crystals more or less modified on the edges, also granular and massive, sp. gr. 4.85 to 5.12, slightly magnetic. Color, black with metallic lustre. Dark-brown powder. Resembles magnetite. Its composition is variable, the average being 66 per cent. of red oxide of iron. The proportions are: sesquioxide of iron 66 to 69 per cent.; oxide of manganese 7.2 to 18 per cent.; and oxide of zinc 11 to 25 per cent. When crystallized, pure, and undecomposed, no chlorine is produced by the treatment with hydrochloric acid, which indicates that the manganese is in the state of protoxide. Ebelmen, after having succeeded in producing a ferrite of zinc, possessing all the properties of Franklinite, thought that this ore was a double ferrite of zinc and manganese.

Franklinite is a valuable iron ore, after its zinc has been separated. It is found in several localities, but the largest deposit is in Sussex Co., New Jersey, between Franklin and Sparta, in granular limestone, accompanied by garnets and an oxide of zinc colored red by an oxide of manganese.

Chamoisite, Berthierite. These are hydrated silicates of alumina, and protoxide of iron, more or less magnetic, the more magnetic being the Berthierite, which is bluish-gray to black, while Chamoisite is greenish-gray to black. The proportion of protoxide of iron varies between 60 to 75 per cent. Chamoisite occurs in Switzerland, and Berthierite in Champagne, in Burgundy, etc., in ammonite limestone. The structure of the ore is compact or oolitic.

ATOMIC WEIGHTS OF ELEMENTARY BODIES.

Aluminium . . .	Al 27.4	Molybdenum . . .	Mo 96.0
Antimony	Sb 122.0	Nickel	Ni 58.7
Arsenic	As 75.0	Niobium	Nb ?
Barium	Ba 137.0	Nitrogen	N 14.0
Bismuth	Bi 210.0	Norium	No ?
Boron	Bo 11.0	Osmium	Os 199.2
Bromine	Br 80.0	Oxygen	O 16.0
Cadmium	Cd 112.0	Palladium	Pa 106.6
Cæsium	Cs 133.0	Phosphorus . . .	Ph 31.0
Calcium	Ca 40.0	Platinum	Pt 197.5
Carbon	C 12.0	Potassium	K 39.1
Cerium	Ce 92.0	Rhodium	Rh 104.4
Chlorine	Cl 35.5	Rubidium	Rb 85.4
Chromium . . .	Cr 52.5	Ruthenium . . .	Ru 104.4
Cobalt	Co 58.7	Selenium	Se 79.5
Copper	Cu 63.5	Silicon	Si 28.0
Didymium . . .	Di 96.0	Silver	Ag 108.0
Erbium	E ?	Sodium	Na 23.0
Fluorine	Fl 19.0	Strontium	Sr 87.5
Glucinum	Gl 9.3	Sulphur	S 32.0
Gold	Au 197.0	Tantalum	Ta 137.6
Hydrogen	H 1.0	Tellurium	Te 129.0
Indium	In 74.0	Thallium	Tl 203.0
Iodine	I 126.8	Thorium	Th 231.5
Iridium	Ir 198.0	Tin	Sn 118.0
Iron	Fe 56.0	Titanium	Ti 50.0
Lanthanum . . .	La 92.0	Tungsten	W 187.0
Lead	Pb 207.0	Uranium	U 120.0
Lithium	Li 7.0	Vanadium . . .	V 137.0
Magnesium . . .	Mg 24.0	Yttrium	Y ?
Manganese . . .	Mn 55.0	Zinc	Zn 65.2
Mercury	Hg 200.0	Zirconium . . .	Zr 89.6

TABLE FOR FACILITATING THE CALCULATION OF ANALYSES.

By multiplying any weight whatsoever of	By the coefficient	We obtain the corresponding weight of	By multiplying any weight whatsoever of	By the coefficient	We obtain the corresponding weight of
Al^2O^3	0.53398	Al	Fe^2O^3	0.90000	Fe^2O^2
As^2O^5	0.65217	As	$Mg^2Ph^2O^7$	0.36036	MgO
$BaSO^4$	0.58369	$CaSO^4$	"	0.27926	Ph
"	0.25751	FeS^2	"	0.63964	Ph^2O^5
"	0.13734	S	Mn^3O^4	0.72052	Mn
"	0.34335	SO^3	$PbCrO^4$	0.31062	CrO^3
CO^2	0.27273	C	$PbSO^4$	0.68317	Pb
CaO	1.78572	$CaCO^3$	"	0.88119	$PbCO^3$
$CaCO^3$	0.56000	CaO	"	0.73597	PbO
$CaSO^4$	0.41176	"	"	0.78878	PbS
Cr^2O^3	0.68619	Cr	Ph	2.29032	Ph^2O^5
"	1.31381	CrO^3	Ph^2O^5	0.43662	Ph
CuO	0.79849	Cu	S	1.87500	FeS^2
Cu^2S	0.79849	"	SiO^2	0.46667	Si
Fe	1.28572	Fe^2O^2	SO^3	1.70000	$CaSO^4$
"	1.42857	Fe^2O^3	"	0.40000	S
"	2.14286	FeS^2	TiO^2	0.6098	Ti
Fe^2O^2	0.77778	Fe	ZnO	0.80260	Zn
"	1.11111	Fe^2O^3	"	1.01970	ZnS
Fe^2O^3	0.70000	Fe	"	1.54187	$ZnCO^3$

TABLES

RELATIVE VALUES OF FRENCH AND ENGLISH WEIGHTS AND MEASURES, &c.

Measures of Length.

Millimetre	=	0.03937	inch.
Centimetre	=	0.393708	"
Decimetre	=	3.937079	inches.
Metre	=	39.37079	"
"	=	3.2808992	feet.
"	=	1.093633	yard.
Decametre	=	32.808992	feet.
Hectometre	=	328.08992	"
Kilometre	=	3280.8992	"
"	=	1093.633	yards.
Myriametre	=	10936.33	"
"	=	6.2138	miles.
Inch ($\frac{1}{36}$ yard)	=	2.539954	centimetres.
Foot ($\frac{1}{3}$ yard)	=	3.0479449	decimetres.
Yard	=	0.91438348	metre.
Fathom (2 yards)	=	1.82876696	"
Pole or perch (5$\frac{1}{2}$ yards)	=	5.029109	metres.
Furlong (220 yards)	=	201.16437	"
Mile (1760 yards)	=	1609.3149	"
Nautical mile	=	1852	"

23   1

Superficial Measures.

Square millimetre		$=$	$\frac{1}{645}$	square inch.	
"	"	$=$	0.00155	"	"
"	centimetre	$=$	0.155006	"	"
"	decimetre	$=$	15.50059	"	inches.
"	"	$=$	0.107643	"	foot.
"	metre or centiare	$=$	1550.05989	"	inches.
"	" "	$=$	10.764299	"	feet.
"	" "	$=$	1.196033	"	yard
Are		$=$	1076.4299	"	feet.
"		$=$	119.6033	"	yards.
"		$=$	0.098845	rood.	
Hectare		$=$	11960.3326	square yards.	
"		$=$	2.471143	acres.	
Square inch		$=$	645.109201	square millimetres.	
" "		$=$	6.451367	"	centimetres
"	foot	$=$	9.289968	"	decimetres.
"	yard	$=$	0.836097	"	metre.
"	rod or perch	$=$	25.291939	"	metres.
Rood (1210 sq. yards)		$=$	10.116775	ares.	
Acre (4840 sq. yards)		$=$	0.404671	hectare.	

Measures of Capacity.

Cubic millimetre		$=$	0.000061027	cubic inch.
"	centimetre or millilitre	$=$	0.061027	" "
10 "	centimetres or centilitre	$=$	0.61027	" "
100 "	" " decilitre	$=$	6.102705	" inches.
1000 "	" " litre	$=$	61.0270515	" "
" "	" " "	$=$	1.760773	imp'l pint.
" "	" " "	$=$	0.2200967	" gal'n.
Decalitre		$=$	610.270515	cubic inches.
"		$=$	2.2009668	imp. gal'ns.
Hectolitre		$=$	3.531658	cubic feet.
"		$=$	22.009668	imp. gal'ns.
Cubic metre or stere or kilolitre		$=$	1.30802	cubic yard.
" " "		$=$	35.3165807	" feet.
Myrialitre		$=$	353.165807	" "

2

Cubic inch	=	16.386176	cubic centimetres.
" foot	=:	28.315312	" decimetres.
" yard	=	0.764513422	" metre.

American Measures.

Winchester or U.S. gallon (231 cub. in.)	=	3.785209 litres.
" " bushel(2150.42 cub. in.)=		35.23719 "
Chaldron (57.25 cubic feet)		= 1621.085 "

British Imperial Measures.

Gill	= 0.141983	litre.
Pint ($\frac{1}{8}$ gallon)	= 0.567932	"
Quart ($\frac{1}{4}$ gallon)	= 1.135864	"
Imperial gallon (277.2738 cub. in.) =	4.54345797	litres.
Peck (2 gallons)	= 9.0869159	"
Bushel (8 gallons)	= 36.347664	"
Sack (3 bushels)	= 1.09043	hectolitre.
Quarter (8 bushels)	= 2.907813	hectolitres.
Chaldron (12 sacks)	= 13.08516	"

Weights.

Milligramme	=	0.015438395	troy grain.
Centigramme	=	0.15438395	" "
Decigramme	=	1.5438395	" "
Gramme	=	15.438395	" grains.
"	=	0.643	pennyweight.
"	=	0.0321633	oz. troy.
"	=	0.0352889	oz. avoirdupois.
Decagramme	=	154.38395	troy grains.
"	=	5.64	drachms avoirdupois.
Hectogramme	=	3.21633	oz. troy.
"	=	3.52889	oz. avoirdupois.
Kilogramme	=	2.6803	lbs. troy.
"	=	2.205486	lbs. avoirdupois.
Myriagramme	=	26.803	lbs. troy.
"	=	22.05486	lbs. avoirdupois.

Quintal metrique = 100 kilog. = 220.5486 lbs. avoirdupois.
Tonne = 1000 kilog. = 2205.486 " "

3

Different authors give the following values for the gramme:—

Gramme = 15.44402   troy grains.
  "   = 15.44242   "
  "   = 15.4402   "
  "   = 15.433159   "
  "   = 15.43234874 "

AVOIRDUPOIS.

Long ton = 20 cwt. = 2240 lbs. = 1015.649	kilogrammes.
Short ton (2000 lbs.) = 906.8296	"
Hundred weight (112 lbs.) = 50.78245	"
Quarter (28 lbs.) = 12.6956144	"
Pound = 16 oz. = 7000 grs. = 453.4148	grammes.
Ounce = 16 dr'ms. = 437.5 grs. = 28.3375	"
Drachm = 27.344 grains = 1.77108	gramme.

TROY (PRECIOUS METALS).

Pound = 12 oz. = 5760 grs. = 373.096	grammes.
Ounce = 20 dwt. = 480 grs. = 31.0913	"
Pennyweight = 24 grs. = 1.55457	gramme.
Grain = 0.064773	"

APOTHECARIES' (PHARMACY).

Ounce = 8 drachms = 480 grs. = 31.0913	gramme.
Drachm = 3 scruples = 60 grs. = 3.8869	"
Scruple = 20 grs. = 1.29546	gramme.

CARAT WEIGHT FOR DIAMONDS.

1 carat = 4 carat grains = 64 carat parts.
  "   = 3.2   troy grains.
  "   = 3.273 "   "
  "   = 0.207264 gramme
  "   = 0.212   "
  "   = 0.205   "

Great diversity in value.

4

Proposed Symbols for Abbreviations.

		Mm	Mg	Ml	
M—myria	— 10000	Mm	Mg	Ml	
K—kilo	— 1000	Km	Kg	Kl	
H—hecto	— 100	Hm	Hg	Hl	Ha
D—deca	— 10	Dm	Dg	Dl	Da
Unit	— 1	metre—m	gramme—g	litre—l	are—a
d—deci	— 0.1	dm	dg	dl	da
c—centi	— 0.01	cm	cg	cl	ca
m—milli	— 0.001	mm	mg	ml	

Km = Kilometre.   Hl = Hectolitre.   cg = centigramme.
c. cm = $\overline{cm}^3$ = cubic centimetre. $\overline{dm}^2$ = sq. dm = square decimetre.   Kgm = Kilogrammetre.   Kg° = Kilogramme degree.

Celsius or Centigrade.	Fahrenheit.	Réaumur.
— 15°	+ 5°	— 12°
— 10	+ 14	— 8
— 5	+ 23	— 4
0 melting	+ 32	ice 0
+ 5	+ 41	+ 4
+ 10	+ 50	+ 8
+ 15	+ 59	+ 12
+ 20	+ 68	+ 16
+ 25	+ 77	+ 20
+ 30	+ 86	+ 24
+ 35	+ 95	+ 28
+ 40	+104	+ 32
+ 45	+113	+ 36
+ 50	+122	+ 40
+ 55	+131	+ 44
+ 60	+140	+ 48
+ 65	+149	+ 52
+ 70	+158	+ 56
+ 75	+167	+ 60
+ 80	+176	+ 64
+ 85	+185	+ 68
+ 90	+194	+ 72
+ 95	+203	+ 76
+100 boiling	+212	water + 80
+200	+392	+160
+300	+572	+240
+400	+752	+320
+500	+932	+400

$$1° \text{ C.} = 1°.8 \text{ Ft.} = \tfrac{9}{5}° \text{ Ft.} = 0°.3 \text{ R.} = \tfrac{4}{5}° \text{ R.}$$

$$1° \text{ C.} \times \tfrac{9}{5} = 1° \text{ Ft.} \qquad 1° \text{ Ft.} \times \tfrac{5}{9} = 1° \text{ C.} \qquad 1° \text{ R.} \times \tfrac{9}{4} = 1° \text{ Ft.}$$

$$1° \text{ C.} \times \tfrac{4}{5} = 1° \text{ R.} \qquad 1° \text{ Ft.} \times \tfrac{4}{9} = 1° \text{ R.} \qquad 1° \text{ R.} \times \tfrac{5}{4} = 1° \text{ C.}$$

Calorie (French) = unit of heat
              = kilogramme degree $\Big\}$ English.

It is the quantity of heat necessary to raise 1° C. the temperature of 1 kilogramme of distilled water.

Kilogrammetre = Kgm = the power necessary to raise 1 kilogramme, 1 metre high, in one second. It is equal to $\frac{1}{75}$ of a French horse power. An English horse power = 550 foot pounds, while a French horse power = 542.7 foot pounds.

Ready-made Calculations.

No. of units.	Inches to centimetres.	Feet to metres.	Yards to metres.	Miles to Kilometres.	Millimetres to inches.
1	2.53995	0.3047945	0.91438348	1.6093	0.03937079
2	5.0799	0.6095890	1.82876696	3.2186	0.07874158
3	7.6199	0.9143835	2.74315044	4.8279	0.11811237
4	10.1598	1.2197680	3.65753392	6.4373	0.15748316
5	12.6998	1.5239724	4.57191740	8.0466	0.19685395
6	15.2397	1.8287669	5.48630088	9.6559	0.23622474
7	17.7797	2.1335614	6.40068436	11.2652	0.27559553
8	20.3196	2.4383559	7.31506784	12.8745	0.31496632
9	22.8596	2.7431504	8.22945132	14.4838	0.35433711
10	25.3995	3.0479450	9.14383480	16.0930	0.39370790

No. of units.	Centimetres to inches.	Metres to feet.	Metres to yards.	Kilometres to miles.	Square inches to square centimetres.
1	0.3937079	3.2808992	1.093633	0.6213824	6.45136
2	0.7874158	6.5617984	2.187266	1.2427648	12.90272
3	1.1811237	9.8426976	3.280899	1.8641472	19.35408
4	1.5748316	13.1235968	4.374532	2.4855296	25.80544
5	1.9685395	16.4044960	5.468165	3.1069120	32.25680
6	2.3622474	19.6853952	6.561798	3.7282944	38.70816
7	2.7559553	22.9662944	7.655431	4.3496768	45.15952
8	3.1496632	26.2471936	8.749064	4.9710592	51.61088
9	3.5433711	29.5280928	9.842697	5.5924416	58.06224
10	3.9370790	32.8089920	10.936330	6.2138240	64.51360

6

No. of units.	Square feet to sq. metres.	Sq. yards to sq. metres.	Acres to hectares.	Square centimetres to sq. inches.	Sq. metres to sq. feet.
1	0.0929	0.836097	0.404671	0.155	10.7643
2	0.1858	1.672194	0.809342	0.310	21.5286
3	0.2787	2.508291	1.204013	0.465	32.2929
4	0.3716	3.344388	1.618684	0.620	43.0572
5	0.4645	4.180485	2.023355	0.775	53.8215
6	0.5574	5.016582	2.428026	0.930	64.5858
7	0.6503	5.852679	2.832697	1.085	75.3501
8	0.7432	6.688776	3.237368	1.240	86.1144
9	0.8361	7.524873	3.642039	1.395	96.8787
10	0.9290	8.360970	4.046710	1.550	107.6430

No. of units.	Square metres to sq. yards.	Hectares to acres.	Cubic inches to cubic centimetres.	Cubic feet to cubic metres.	Cubic yards to cubic metres.
1	1.196033	2.471143	16.3855	0.02831	0.76451
2	2.392066	4.942286	32.7710	0.05662	1.52902
3	3.588099	7.413429	49.1565	0.08494	2.29354
4	4.784132	9.884572	65.5420	0.11325	3.05805
5	5.980165	12.355715	81.9275	0.14157	3.82257
6	7.176198	14.826858	98.3130	0.16988	4.58708
7	8.372231	17.298001	114.6985	0.19819	5.35159
8	9.568264	19.769144	131.0840	0.22651	6.11611
9	10.764297	22.240287	147.4695	0.25482	6.88062
10	11.960330	24.711430	163.8550	0.28315	7.64513

No. of units.	Cubic centimetres to cubic inches.	Litres to cubic inches.	Hectolitres to cubic feet.	Cubic metres to cubic feet.	Cubic metres to cubic yards.
1	0.06102	61.02705	3.5317	35.31659	1.30802
2	0.12205	122.05410	7.0634	70.63318	2.61604
3	0.18308	183.08115	10.5951	105.94977	3.92406
4	0.24411	244.10820	14.1268	141.26636	5.23208
5	0.30514	305.13525	17.6585	176.58295	6.54010
6	0.36617	366.16230	21.1902	211.89954	7.84812
7	0.42720	427.18935	24.7219	247.21613	9.15614
8	0.48823	488.21640	28.2536	282.53272	10.46416
9	0.54926	549.24345	31.7853	317.84931	11.77218
10	0.61027	610.27050	35.3166	353.16590	13 08020

No. of units.	Grains to grammes.	Ounces avoir. to grammes.	Ounces troy to grammes.	Pounds avoir. to kilogrammes.	Pounds troy to kilogrammes.
1	0.064773	28.3375	31.0913	0.4534148	0.373096
2	0.129546	56.6750	62.1826	0.9068296	0.746192
3	0.194319	85.0125	93.2739	1.3602444	1.119288
4	0.259092	113.3500	124.3652	1.8136592	1.492384
5	0.323865	141.6871	155.4565	2.2670740	1.865480
6	0.388638	170.0250	186.5478	2.7204888	2.238576
7	0.453411	198.3625	217.6391	3.1739036	2.611672
8	0.518184	226.7000	248.7304	3.6273184	2.984768
9	0.582957	255.0375	279.8217	4.0807332	3.357864
10	0.647730	283.3750	310.9130	4.5341480	3.730960

No. of units.	Long tons to tonnes of 1000 kilog.	Pounds per square inch to kilogrammes per square centimetre.	Grammes to grains.	Grammes to ounces avoir.	Grammes to ounces troy.
1	1.015649	0.0702774	15.438395	0.0352889	0.0321633
2	2.031298	0.1405548	30.876790	0.0705778	0.0643266
3	3.046947	0.2108322	46.315185	0.1058667	0.0964899
4	4.062596	0.2811096	61.753580	0.1411556	0.1286532
5	5.078245	0.3513870	77.191975	0.1764445	0.1608165
6	6.093894	0.4216644	92.630370	0.2117334	0.1929798
7	7.109543	0.4919418	108.068765	0.2470223	0.2251431
8	8.125192	0.5622192	123.507160	0.2823112	0.2573064
9	9.140841	0.6324966	138.945555	0.3176001	0.2894697
10	10.156490	0.7027740	154.383950	0.3528890	0.3216330

No. of units.	Kilogrammes to pounds avoirdupois.	Kilogrammes to pounds troy.	Metric tonnes of 1000 kilog. to long tons of 2240 pounds.	Kilog. per square millimetre to pounds per square inch.	Kilog. per square centimetre to pounds per square inch.
1	2.205486	2.6803	0.9845919	1422.52	14.22526
2	4.410972	5.3606	1.9691838	2845.05	28.45052
3	6.616458	8.0409	2.9537757	4267.57	42.67578
4	8.821944	10.7212	3.9383676	5690.10	56.90104
5	11.027430	13.4015	4.9229595	7112.63	71.12630
6	13.232916	16.0818	5.9075514	8535.15	85.35156
7	15.438402	18.7621	6.8921433	9957.68	99.57682
8	17.643888	21.4424	7.8767352	11380.20	113.80208
9	19.849374	24.1227	8.8613271	12802.73	128.02734
10	22.054860	26.8030	9.8459190	14225.26	142.25260

INDEX.

24

CATALOGUE

OF

PRACTICAL AND SCIENTIFIC BOOKS,

PUBLISHED BY

HENRY CAREY BAIRD,

Industrial Publisher,

NO. 406 WALNUT STREET,

PHILADELPHIA.

☞ Any of the Books comprised in this Catalogue will be sent by mail, free of postage, at the publication price.

☞ A Descriptive Catalogue, 96 pages, 8vo., will be sent, free of postage, to any one who will furnish the publisher with his address.

ARLOT.—A Complete Guide for Coach Painters.

Translated from the French of M. ARLOT, Coach Painter; for eleven years Foreman of Painting to M. Eherler, Coach Maker, Paris. By A. A. FESQUET, Chemist and Engineer. To which is added an Appendix, containing Information respecting the Materials and the Practice of Coach and Car Painting and Varnishing in the United States and Great Britain. 12mo. $1.25

ARMENGAUD, AMOROUX, and JOHNSON.—The Practical Draughtsman's Book of Industrial Design, and Machinist's and Engineer's Drawing Companion:

Forming a Complete Course of Mechanical Engineering and Architectural Drawing. From the French of M. Armengaud the elder, Prof. of Design in the Conservatoire of Arts and Industry, Paris, and MM. Armengaud the younger, and Amoroux, Civil Engineers. Rewritten and arranged with additional matter and plates, selections from and examples of the most useful and generally employed mechanism of the day. By WILLIAM JOHNSON, Assoc. Inst. C. E., Editor of "The Practical Mechanic's Journal." Illustrated by 50 folio steel plates, and 50 wood-cuts. A new edition, 4to. $10.00

1

ARROWSMITH.—Paper-Hanger's Companion:

A Treatise in which the Practical Operations of the Trade are Systematically laid down: with Copious Directions Preparatory to Papering; Preventives against the Effect of Damp on Walls; the Various Cements and Pastes Adapted to the Several Purposes of the Trade; Observations and Directions for the Panelling and Ornamenting of Rooms, etc. By JAMES ARROWSMITH, Author of "Analysis of Drapery," etc. 12mo., cloth. $1.25

ASHTON.—The Theory and Practice of the Art of Designing Fancy Cotton and Woollen Cloths from Sample:

Giving full Instructions for Reducing Drafts, as well as the Methods of Spooling and Making out Harness for Cross Drafts, and Finding any Required Reed, with Calculations and Tables of Yarn. By FREDERICK T. ASHTON, Designer, West Pittsfield, Mass. With 52 Illustrations. One volume, 4to. $10.00

BAIRD.—Letters on the Crisis, the Currency and the Credit System.

By HENRY CAREY BAIRD. Pamphlet.   05

BAIRD.—Protection of Home Labor and Home Productions necessary to the Prosperity of the American Farmer.

By HENRY CAREY BAIRD. 8vo., paper.   10

BAIRD.—Some of the Fallacies of British Free-Trade Revenue Reform.

Two Letters to Arthur Latham Perry, Professor of History and Political Economy in Williams College. By HENRY CAREY BAIRD. Pamphlet.   05

BAIRD.—The Rights of American Producers, and the Wrongs of British Free-Trade Revenue Reform.

By HENRY CAREY BAIRD. Pamphlet.   05

BAIRD.—Standard Wages Computing Tables:

An Improvement in all former Methods of Computation, so arranged that wages for days, hours, or fractions of hours, at a specified rate per day or hour, may be ascertained at a glance. By T. SPANGLER BAIRD. Oblong folio. $5.00

BAIRD.—The American Cotton Spinner, and Manager's and Carder's Guide:

A Practical Treatise on Cotton Spinning; giving the Dimensions and Speed of Machinery, Draught and Twist Calculations, etc.; with notices of recent Improvements: together with Rules and Examples for making changes in the sizes and numbers of Roving and Yarn. Compiled from the papers of the late ROBERT H. BAIRD. 12mo. $1.50

BAKER.—Long-Span Railway Bridges :

Comprising Investigations of the Comparative Theoretical and Practical Advantages of the various Adopted or Proposed Type Systems of Construction ; with numerous Formulæ and Tables. By B. BAKER. 12mo.　.　.　.　.　.　.　.　.　.　.　$2.00

BAUERMAN.—A Treatise on the Metallurgy of Iron :

Containing Outlines of the History of Iron Manufacture, Methods of Assay, and Analysis of Iron Ores, Processes of Manufacture of Iron and Steel, etc., etc. By H. BAUERMAN, F. G. S., Associate of the Royal School of Mines. First American Edition, Revised and Enlarged. With an Appendix on the Martin Process for Making Steel, from the Report of ABRAM S. HEWITT, U. S. Commissioner to the Universal Exposition at Paris, 1867. Illustrated. 12mo.　.　$2.00

BEANS.—A Treatise on Railway Curves and the Location of Railways.

By E. W. BEANS, C. E. Illustrated. 12mo. Tucks.　.　.　$1.50

BELL.—Carpentry Made Easy :

Or, The Science and Art of Framing on a New and Improved System. With Specific Instructions for Building Balloon Frames, Barn Frames, Mill Frames, Warehouses, Church Spires, etc. Comprising also a System of Bridge Building, with Bills, Estimates of Cost, and valuable Tables. Illustrated by 38 plates, comprising nearly 200 figures. By WILLIAM E. BELL, Architect and Practical Builder. 8vo.　.　$5.00

BELL.—Chemical Phenomena of Iron Smelting :

An Experimental and Practical Examination of the Circumstances which determine the Capacity of the Blast Furnace, the Temperature of the Air, and the proper Condition of the Materials to be operated upon. By I. LOWTHIAN BELL. Illustrated. 8vo.　.　.　$6.00

BEMROSE.—Manual of Wood Carving :

With Practical Illustrations for Learners of the Art, and Original and Selected Designs. By WILLIAM BEMROSE, Jr. With an Introduction by LLEWELLYN JEWITT, F. S. A., etc. With 128 Illustrations. 4to., cloth.　.　.　.　.　.　.　.　.　.　.　$3.00

BICKNELL.—Village Builder, and Supplement :

Elevations and Plans for Cottages, Villas, Suburban Residences, Farm Houses, Stables and Carriage Houses. Store Fronts, School Houses, Churches, Court Houses, and a model Jail ; also, Exterior and Interior details for Public and Private Buildings, with approved Forms of Contracts and Specifications, including Prices of Building Materials and Labor at Boston, Mass., and St. Louis, Mo. Containing 75 plates drawn to scale; showing the style and cost of building in different sections of the country, being an original work comprising the designs of twenty leading architects, representing the New England, Middle, Western, and Southwestern States. 4to.　.　$12.00

BLENKARN.—Practical Specifications of Works executed in Architecture, Civil and Mechanical Engineering, and in Road Making and Sewering :

To which are added a series of practically useful Agreements and Reports. By JOHN BLENKARN. Illustrated by 15 large folding plates. 8vo. $9.00

BLINN.—A Practical Workshop Companion for Tin, Sheet-Iron, and Copperplate Workers :

Containing Rules for describing various kinds of Patterns used by Tin, Sheet-Iron, and Copper-plate Workers; Practical Geometry; Mensuration of Surfaces and Solids; Tables of the Weights of Metals, Lead Pipe, etc.; Tables of Areas and Circumferences of Circles; Japan, Varnishes, Lackers, Cements, Compositions, etc., etc. By LEROY J. BLINN, Master Mechanic. With over 100 Illustrations. 12mo. $2.50

BOOTH.—Marble Worker's Manual :

Containing Practical Information respecting Marbles in general, their Cutting, Working, and Polishing; Veneering of Marble; Mosaics; Composition and Use of Artificial Marble, Stuccos, Cements, Receipts, Secrets, etc., etc. Translated from the French by M. L. BOOTH. With an Appendix concerning American Marbles. 12mo., cloth. $1.50

BOOTH AND MORFIT.—The Encyclopedia of Chemistry, Practical and Theoretical :

Embracing its application to the Arts, Metallurgy, Mineralogy, Geology, Medicine, and Pharmacy. By JAMES C. BOOTH, Melter and Refiner in the United States Mint, Professor of Applied Chemistry in the Franklin Institute, etc., assisted by CAMPBELL MORFIT, author of "Chemical Manipulations," etc. Seventh edition. Royal 8vo., 978 pages, with numerous wood-cuts and other illustrations. . $5.00

BOX.—A Practical Treatise on Heat :

As applied to the Useful Arts; for the Use of Engineers, Architects, etc. By THOMAS BOX, author of "Practical Hydraulics." Illustrated by 14 plates containing 114 figures. 12mo. $4.25

BOX.—Practical Hydraulics :

A Series of Rules and Tables for the use of Engineers, etc. By THOMAS BOX. 12mo. $2.50

BROWN.—Five Hundred and Seven Mechanical Movements :

Embracing all those which are most important in Dynamics, Hydraulics, Hydrostatics, Pneumatics, Steam Engines, Mill and other Gearing, Presses, Horology, and Miscellaneous Machinery; and including many movements never before published, and several of which have only recently come into use. By HENRY T. BROWN, Editor of the "American Artisan." In one volume, 12mo. . . . $1.00

BUCKMASTER.—The Elements of Mechanical Physics:

By J. C. BUCKMASTER, late Student in the Government School of Mines; Certified Teacher of Science by the Department of Science and Art; Examiner in Chemistry and Physics in the Royal College of Preceptors; and late Lecturer in Chemistry and Physics of the Royal Polytechnic Institute. Illustrated with numerous engravings. In one volume, 12mo. $1.50

BULLOCK.—The American Cottage Builder:

A Series of Designs, Plans, and Specifications, from $200 to $20,000, for Homes for the People; together with Warming, Ventilation, Drainage, Painting, and Landscape Gardening. By JOHN BULLOCK, Architect, Civil Engineer, Mechanician, and Editor of "The Rudiments of Architecture and Building," etc., etc. Illustrated by 75 engravings. In one volume, 8vo. $3.50

BULLOCK. — The Rudiments of Architecture and Building:

For the use of Architects, Builders, Draughtsmen, Machinists, Engineers, and Mechanics. Edited by JOHN BULLOCK, author of "The American Cottage Builder." Illustrated by 250 engravings. In one volume, 8vo. $3.50

BURGH.—Practical Illustrations of Land and Marine Engines:

Showing in detail the Modern Improvements of High and Low Pressure, Surface Condensation, and Super-heating, together with Land and Marine Boilers. By N. P. BURGH, Engineer. Illustrated by 20 plates, double elephant folio, with text.   .   .   . $21.00

BURGH.—Practical Rules for the Proportions of Modern Engines and Boilers for Land and Marine Purposes.

By N. P. BURGH, Engineer. 12mo. $1.50

BURGH.—The Slide-Valve Practically Considered.

By N. P. BURGH, Engineer. Completely illustrated. 12mo. $2.00

BYLES.—Sophisms of Free Trade and Popular Political Economy Examined.

By a BARRISTER (Sir JOHN BARNARD BYLES, Judge of Common Pleas). First American from the Ninth English Edition, as published by the Manchester Reciprocity Association. In one volume, 12mo. Paper, 75 cts. Cloth $1.25

BYRN.—The Complete Practical Brewer:

Or Plain, Accurate, and Thorough Instructions in the Art of Brewing Beer, Ale, Porter, including the Process of making Bavarian Beer, all the Small Beers, such as Root-beer, Ginger-pop, Sarsaparilla-beer, Mead, Spruce Beer, etc., etc. Adapted to the use of Public Brewers and Private Families. By M. LA FAYETTE BYRN, M. D. With illustrations. 12mo. $1.25

BYRN.—The Complete Practical Distiller:

Comprising the most perfect and exact Theoretical and Practical Description of the Art of Distillation and Rectification; including all of the most recent improvements in distilling apparatus; instructions for preparing spirits from the numerous vegetables, fruits, etc.; directions for the distillation and preparation of all kinds of brandies and other spirits, spirituous and other compounds, etc., etc. By M. LA FAYETTE BYRN, M. D. Eighth Edition. To which are added, Practical Directions for Distilling, from the French of Th. Fling, Brewer and Distiller. 12mo. $1.50

BYRNE.—Handbook for the Artisan, Mechanic, and Engineer:

Comprising the Grinding and Sharpening of Cutting Tools, Abrasive Processes, Lapidary Work, Gem and Glass Engraving, Varnishing and Lackering, Apparatus, Materials and Processes for Grinding and Polishing, etc. By OLIVER BYRNE. Illustrated by 185 wood engravings. In one volume, 8vo. $5.00

BYRNE.—Pocket Book for Railroad and Civil Engineers:

Containing New, Exact, and Concise Methods for Laying out Railroad Curves, Switches, Frog Angles, and Crossings; the Staking out of work; Levelling; the Calculation of Cuttings; Embankments; Earth-work, etc. By OLIVER BYRNE. 18mo., full bound, pocketbook form. $1.75

BYRNE.—The Practical Model Calculator:

For the Engineer, Mechanic, Manufacturer of Engine Work, Naval Architect, Miner, and Millwright. By OLIVER BYRNE. 1 volume, 8vo., nearly 600 pages $4.50

BYRNE.—The Practical Metal-Worker's Assistant:

Comprising Metallurgic Chemistry; the Arts of Working all Metals and Alloys; Forging of Iron and Steel; Hardening and Tempering; Melting and Mixing; Casting and Founding; Works in Sheet Metal; The Processes Dependent on the Ductility of the Metals; Soldering; and the most Improved Processes and Tools employed by Metal-Workers. With the Application of the Art of Electro-Metallurgy to Manufacturing Processes; collected from Original Sources, and from the Works of Holtzapffel, Bergeron, Leupold, Plumier, Napier, Scoffern, Clay, Fairbairn, and others. By OLIVER BYRNE. A new, revised, and improved edition, to which is added An Appendix, containing THE MANUFACTURE OF RUSSIAN SHEET-IRON. By JOHN PERCY, M. D., F.R.S. THE MANUFACTURE OF MALLEABLE IRON CASTINGS, and IMPROVEMENTS IN BESSEMER STEEL. By A. A. FESQUET, Chemist and Engineer. With over 600 Engravings, illustrating every Branch of the Subject. 8vo. $7.00

Cabinet Maker's Album of Furniture:

Comprising a Collection of Designs for Furniture. Illustrated by 48 Large and Beautifully Engraved Plates. In one vol., oblong $5.00

CALLINGHAM.—Sign Writing and Glass Embossing:

A Complete Practical Illustrated Manual of the Art. By JAMES
CALLINGHAM. In one volume, 12mo. $1.50

CAMPIN.—A Practical Treatise on Mechanical Engineering:

Comprising Metallurgy, Moulding, Casting, Forging, Tools, Workshop Machinery, Mechanical Manipulation, Manufacture of Steamengines, etc., etc. With an Appendix on the Analysis of Iron and Iron Ores. By FRANCIS CAMPIN, C. E. To which are added, Observations on the Construction of Steam Boilers, and Remarks upon Furnaces used for Smoke Prevention; with a Chapter on Explosions. By R. Armstrong, C. E., and John Bourne. Rules for Calculating the Change Wheels for Screws on a Turning Lathe, and for a Wheelcutting Machine. By J. LA NICCA. Management of Steel, Including Forging, Hardening, Tempering, Annealing, Shrinking, and Expansion. And the Case-hardening of Iron. By G. EDE. 8vo. Illustrated with 29 plates and 100 wood engravings . . . $6.00

CAMPIN.—The Practice of Hand-Turning in Wood, Ivory, Shell, etc.:

With Instructions for Turning such works in Metal as may be required in the Practice of Turning Wood, Ivory, etc. Also, an Appendix on Ornamental Turning. By FRANCIS CAMPIN; with Numerous Illustrations. 12mo., cloth $3.00

CAREY.—The Works of Henry C. Carey:

FINANCIAL CRISES, their Causes and Effects. 8vo. paper . 25

HARMONY OF INTERESTS: Agricultural, Manufacturing, and Commercial. 8vo., cloth $1.50

MANUAL OF SOCIAL SCIENCE. Condensed from Carey's "Principles of Social Science." By KATE MCKEAN. 1 vol. 12mo. $2.25

MISCELLANEOUS WORKS: comprising "Harmony of Interests," "Money," "Letters to the President," "Financial Crises," "The Way to Outdo England Without Fighting Her," "Resources of the Union," "The Public Debt," "Contraction or Expansion?" "Review of the Decade 1857-'67," "Reconstruction," etc., etc. Two vols., 8vo., cloth $10.00

PAST, PRESENT, AND FUTURE. 8vo. $2.50

PRINCIPLES OF SOCIAL SCIENCE. 3 vols., 8vo., cloth $10.00

THE SLAVE-TRADE, DOMESTIC AND FOREIGN; Why it Exists, and How it may be Extinguished (1853). 8vo., cloth . $2.00

LETTERS ON INTERNATIONAL COPYRIGHT (1867) . 50

THE UNITY OF LAW: As Exhibited in the Relations of Physical, Social, Mental, and Moral Science (1872). In one volume, 8vo., pp. xxiii., 433. Cloth $3.50

CHAPMAN.—A Treatise on Ropemaking: ·

As Practised in private and public Rope yards, with a Description of the Manufacture, Rules, Tables of Weights, etc., adapted to the Trades, Shipping, Mining, Railways, Builders, etc. By ROBERT CHAPMAN. 24mo. $1.50

COLBURN.—The Locomotive Engine:

Including a Description of its Structure, Rules for Estimating its Capabilities, and Practical Observations on its Construction and Management. By ZERAH COLBURN. Illustrated. A new edition. 12mo. $1.25

CRAIK. — The Practical American Millwright and Miller.

By DAVID CRAIK, Millwright. Illustrated by numerous wood engravings, and two folding plates. 8vo. $5.00

DE GRAFF.—The Geometrical Stair Builders' Guide:

Being a Plain Practical System of Hand-Railing, embracing all its necessary Details, and Geometrically Illustrated by 22 Steel Engravings; together with the use of the most approved principles of Practical Geometry. By SIMON DE GRAFF, Architect. 4to. . $5.00

DE KONINCK.—DIETZ.—A Practical Manual of Chemical Analysis and Assaying:

As applied to the Manufacture of Iron from its Ores, and to Cast Iron, Wrought Iron, and Steel, as found in Commerce. By L. L. DE KONINCK, Dr. Sc., and E. DIETZ, Engineer. Edited with Notes, by ROBERT MALLET, F.R.S., F.S.G., M.I.C.E., etc. American Edition, Edited with Notes and an Appendix on Iron Ores, by A. A. FESQUET, Chemist and Engineer. One volume, 12mo. $2.50

DUNCAN.—Practical Surveyor's Guide:

Containing the necessary information to make any person, of common capacity, a finished land surveyor without the aid of a teacher. By ANDREW DUNCAN. Illustrated. 12mo., cloth. . . . $1.25

DUPLAIS.—A Treatise on the Manufacture and Distillation of Alcoholic Liquors:

Comprising Accurate and Complete Details in Regard to Alcohol from Wine, Molasses, Beets, Grain, Rice, Potatoes, Sorghum, Asphodel, Fruits, etc.; with the Distillation and Rectification of Brandy, Whiskey, Rum, Gin, Swiss, Absinthe, etc., the Preparation of Aromatic Waters, Volatile Oils or Essences, Sugars, Syrups, Aromatic Tinctures, Liqueurs, Cordial Wines, Effervescing Wines, etc., the Aging of Brandy and the Improvement of Spirits, with Copious Directions and Tables for Testing and Reducing Spirituous Liquors, etc., etc. Translated and Edited from the French of MM. DUPLAIS, Aîné et Jeune. By M. McKENNIE, M.D. To which are added the United States Internal Revenue Regulations for the Assessment and Collection of Taxes on Distilled Spirits. Illustrated by fourteen folding plates and several wood engravings. 743 pp., 8vo. $10.00

DUSSAUCE.—A General Treatise on the Manufacture of Every Description of Soap:

Comprising the Chemistry of the Art, with Remarks on Alkalies, Saponifiable Fatty Bodies, the apparatus necessary in a Soap Factory, Practical Instructions in the manufacture of the various kinds of Soap, the assay of Soaps, etc., etc. Edited from Notes of Larmé, Fontenelle, Malapayre, Dufour, and others, with large and important additions by Prof. H. DUSSAUCE, Chemist. Illustrated. In one vol., 8vo. . $10.00

DUSSAUCE.—A General Treatise on the Manufacture of Vinegar:

Theoretical and Practical. Comprising the various Methods, by the Slow and the Quick Processes, with Alcohol, Wine, Grain, Malt, Cider, Molasses, and Beets; as well as the Fabrication of Wood Vinegar, etc., etc. By Prof. H. DUSSAUCE. In one volume, 8vo. . . $5.00

DUSSAUCE.—A New and Complete Treatise on the Arts of Tanning, Currying, and Leather Dressing:

Comprising all the Discoveries and Improvements made in France, Great Britain, and the United States. Edited from Notes and Documents of Messrs. Sallerou, Grouvelle, Duval, Dessables, Labarraque, Payen, René, De Fontenelle, Malapeyre, etc., etc. By Prof. H. DUSSAUCE, Chemist. Illustrated by 212 wood engravings. 8vo. $20.00

DUSSAUCE.—A Practical Guide for the Perfumer :

Being a New Treatise on Perfumery, the most favorable to the Beauty without being injurious to the Health, comprising a Description of the substances used in Perfumery, the Formulæ of more than 1000 Preparations, such as Cosmetics, Perfumed Oils, Tooth Powders, Waters, Extracts, Tinctures, Infusions, Spirits, Vinaigres, Essential Oils, Pastels, Creams, Soaps, and many new Hygienic Products not hitherto described. Edited from Notes and Documents of Messrs. Debay, Lunel, etc. With additions by Prof. H. DUSSAUCE, Chemist. 12mo. $3.00

DUSSAUCE.—Practical Treatise on the Fabrication of Matches, Gun Cotton, and Fulminating Powders.

By Prof. H. DUSSAUCE. 12mo. $3.00

Dyer and Color-maker's Companion:

Containing upwards of 200 Receipts for making Colors, on the most approved principles, for all the various styles and fabrics now in existence; with the Scouring Process, and plain Directions for Preparing, Washing-off, and Finishing the Goods. In one vol., 12mo. . $1.25

EASTON.—A Practical Treatise on Street or Horse-power Railways.

By ALEXANDER EASTON, C. E. Illustrated by 23 plates. 8vo., cloth. $2.00

ELDER.—Questions of the Day:

Economic and Social. By Dr. WILLIAM ELDER. 8vo. . $3.00

FAIRBAIRN.—The Principles of Mechanism and Machinery of Transmission:

Comprising the Principles of Mechanism, Wheels, and Pulleys, Strength and Proportions of Shafts, Coupling of Shafts, and Engaging and Disengaging Gear. By Sir WILLIAM FAIRBAIRN, C.E., LL.D., F.R.S., F.G.S. Beautifully illustrated by over 150 wood-cuts. In one volume, 12mo. $2.50

FORSYTH.—Book of Designs for Headstones, Mural, and other Monuments:

Containing 78 Designs. By JAMES FORSYTH. With an Introduction by CHARLES BOUTELL, M. A. 4to., cloth. $5.00

GIBSON.—The American Dyer:

A Practical Treatise on the Coloring of Wool, Cotton, Yarn and Cloth, in three parts. Part First gives a descriptive account of the Dye Stuffs; if of vegetable origin, where produced, how cultivated, and how prepared for use; if chemical, their composition, specific gravities, and general adaptability, how adulterated, and how to detect the adulterations, etc. Part Second is devoted to the Coloring of Wool, giving recipes for one hundred and twenty-nine different colors or shades, and is supplied with sixty colored samples of Wool. Part Third is devoted to the Coloring of Raw Cotton or Cotton Waste, for mixing with Wool Colors in the Manufacture of all kinds of Fabrics, gives recipes for thirty-eight different colors or shades, and is supplied with twenty-four colored samples of Cotton Waste. Also, recipes for Coloring Beavers, Doeskins, and Flannels, with remarks upon Anilines, giving recipes for fifteen different colors or shades, and nine samples of Aniline Colors that will stand both the Fulling and Scouring process. Also, recipes for Aniline Colors on Cotton Thread, and recipes for Common Colors on Cotton Yarns. Embracing in all over two hundred recipes for Colors and Shades, and ninety-four samples of Colored Wool and Cotton Waste, etc. By RICHARD H. GIBSON, Practical Dyer and Chemist. In one volume, 8vo. . . $12.50

GILBART.—History and Principles of Banking:

A Practical Treatise. By JAMES W. GILBART, late Manager of the London and Westminster Bank. With additions. In one volume, 8vo., 600 pages, sheep $5.00

Gothic Album for Cabinet Makers:

Comprising a Collection of Designs for Gothic Furniture. Illustrated by 23 large and beautifully engraved plates. Oblong . . $3.00

GRANT. — Beet-root Sugar and Cultivation of the Beet.

By E. B. GRANT. 12mo. $1.25

GREGORY.—Mathematics for Practical Men:

Adapted to the Pursuits of Surveyors, Architects, Mechanics, and Civil Engineers. By OLINTHUS GREGORY. 8vo., plates, cloth $3.00

GRISWOLD.—Railroad Engineer's Pocket Companion for the Field:

Comprising Rules for Calculating Deflection Distances and Angles, Tangential Distances and Angles, and all Necessary Tables for Engineers; also the art of Levelling from Preliminary Survey to the Construction of Railroads, intended Expressly for the Young Engineer, together with Numerous Valuable Rules and Examples. By W. GRISWOLD. 12mo., tucks $1.75

GRUNER.—Studies of Blast Furnace Phenomena.

By M. L. GRUNER, President of the General Council of Mines of France, and lately Professor of Metallurgy at the Ecole des Mines. Translated, with the Author's sanction, with an Appendix, by L. D. B. Gordon, F. R. S. E., F. G. S. Illustrated. 8vo. . . . $2.50

GUETTIER.—Metallic Alloys:

Being a Practical Guide to their Chemical and Physical Properties, their Preparation, Composition, and Uses. Translated from the French of A. GUETTIER, Engineer and Director of Foundries, author of "La Fouderie en France," etc., etc. By A. A. FESQUET, Chemist and Engineer. In one volume, 12mo. $3.00

HARRIS.—Gas Superintendent's Pocket Companion.

By HARRIS & BROTHER, Gas Meter Manufacturers, 1115 and 1117 Cherry Street, Philadelphia. Full bound in pocket-book form $2.00

Hats and Felting:

A Practical Treatise on their Manufacture. By a Practical Hatter. Illustrated by Drawings of Machinery, etc. 8vo. . . . $1.25

HOFMANN.—A Practical Treatise on the Manufacture of Paper in all its Branches.

By CARL HOFMANN. Late Superintendent of paper mills in Germany and the United States; recently manager of the Public Ledger Paper Mills, near Elkton, Md. Illustrated by 110 wood engravings, and five large folding plates. In one volume, 4to., cloth; 398 pages $15.00

HUGHES.—American Miller and Millwright's Assistant.

By WM. CARTER HUGHES. A new edition. In one vol., 12mo. $1.50

HURST.—A Hand-Book for Architectural Surveyors and others engaged in Building:

Containing Formulæ useful in Designing Builder's work, Table of Weights, of the materials used in Building, Memoranda connected with Builders' work, Mensuration, the Practice of Builders' Measurement, Contracts of Labor, Valuation of Property, Summary of the Practice in Dilapidation, etc., etc. By J. F. HURST, C. E. Second edition, pocket-book form, full bound $2.50

JERVIS.—Railway Property:

A Treatise on the Construction and Management of Railways; designed to afford useful knowledge, in the popular style, to the holders of this class of property; as well as Railway Managers, Officers, and Agents. By JOHN B. JERVIS, late Chief Engineer of the Hudson River Railroad, Croton Aqueduct, etc. In one vol., 12mo., cloth $2.00

JOHNSTON.—Instructions for the Analysis of Soils, Limestones, and Manures.

By J. F. W. JOHNSTON. 12mo. 38

KEENE.—A Hand-Book of Practical Gauging:
For the Use of Beginners, to which is added, A Chapter on Distilla‑
tion, describing the process in operation at the Custom House for
ascertaining the strength of wines. By JAMES B. KEENE, of H. M.
Customs. 8vo. $1.25

**KELLEY.—Speeches, Addresses, and Letters on In‑
dustrial and Financial Questions.**
By Hon. WILLIAM D. KELLEY, M. C. In one volume, 544 pages,
8vo. $3.00

KENTISH.—A Treatise on a Box of Instruments,
And the Slide Rule; with the Theory of Trigonometry and Loga‑
rithms, including Practical Geometry, Surveying, Measuring of Tim‑
ber, Cask and Malt Gauging, Heights, and Distances. By THOMAS
KENTISH. In one volume. 12mo. $1.25

KOBELL.—ERNI.—Mineralogy Simplified:
A short Method of Determining and Classifying Minerals, by means
of simple Chemical Experiments in the Wet Way. Translated from
the last German Edition of F. VON KOBELL, with an Introduction to
Blow‑pipe Analysis and other additions. By HENRI ERNI, M. D.,
late Chief Chemist, Department of Agriculture, author of " Coal Oil
and Petroleum." In one volume, 12mo. $2.50

LANDRIN.—A Treatise on Steel:
Comprising its Theory, Metallurgy, Properties, Practical Working,
and Use. By M. H. C. LANDRIN, Jr., Civil Engineer. Translated
from the French, with Notes, by A. A. FESQUET, Chemist and Engi‑
neer. With an Appendix on the Bessemer and the Martin Processes
for Manufacturing Steel, from the Report of Abram S. Hewitt, United
States Commissioner to the Universal Exposition, Paris, 1867. In one
volume, 12mo. $3.00

**LARKIN.—The Practical Brass and Iron Founder's
Guide:**
A Concise Treatise on Brass Founding, Moulding, the Metals and their
Alloys, etc.: to which are added Recent Improvements in the Manu‑
facture of Iron, Steel by the Bessemer Process, etc., etc. By JAMES
LARKIN, late Conductor of the Brass Foundry Department in Reany,
Neafie & Co's. Penn Works, Philadelphia. Fifth edition, revised,
with Extensive additions. In one volume, 12mo. . . $2.25

LEAVITT.—Facts about Peat as an Article of Fuel:
With Remarks upon its Origin and Composition, the Localities in
which it is found, the Methods of Preparation and Manufacture, and
the various Uses to which it is applicable; together with many other
matters of Practical and Scientific Interest. To which is added a chap‑
ter on the Utilization of Coal Dust with Peat for the Production of an
Excellent Fuel at Moderate Cost, specially adapted for Steam Service.
By T. H. LEAVITT. Third edition. 12mo. . . . $1.75

LEROUX, C.—A Practical Treatise on the Manufacture of Worsteds and Carded Yarns:

Comprising Practical Mechanics, with Rules and Calculations applied to Spinning; Sorting, Cleaning, and Scouring Wools; the English and French methods of Combing, Drawing, and Spinning Worsteds and Manufacturing Carded Yarns. Translated from the French of CHARLES LEROUX, Mechanical Engineer, and Superintendent of a Spinning Mill, by HORATIO PAINE, M. D., and A. A. FESQUET, Chemist and Engineer. Illustrated by 12 large Plates. To which is added an Appendix, containing extracts from the Reports of the International Jury, and of the Artisans selected by the Committee appointed by the Council of the Society of Arts, London, on Woollen and Worsted Machinery and Fabrics, as exhibited in the Paris Universal Exposition, 1867. 8vo., cloth. $5.00

LESLIE (Miss).—Complete Cookery:

Directions for Cookery in its Various Branches. By MISS LESLIE. 60th thousand. Thoroughly revised, with the addition of New Receipts. In one volume, 12mo., cloth. $1.50

LESLIE (Miss).—Ladies' House Book:

A Manual of Domestic Economy. 20th revised edition. 12mo., cloth.

LESLIE (Miss).—Two Hundred Receipts in French Cookery.

Cloth, 12mo.

LIEBER.—Assayer's Guide:

Or, Practical Directions to Assayers, Miners, and Smelters, for the Tests and Assays, by Heat and by Wet Processes, for the Ores of all the principal Metals, of Gold and Silver Coins and Alloys, and of Coal, etc. By OSCAR M. LIEBER. 12mo., cloth. . . $1.25

LOTH.—The Practical Stair Builder:

A Complete Treatise on the Art of Building Stairs and Hand-Rails, Designed for Carpenters, Builders, and Stair-Builders. Illustrated with Thirty Original Plates. By C. EDWARD LOTH, Professional Stair-Builder. One large 4to. volume. $10.00

LOVE.—The Art of Dyeing, Cleaning, Scouring, and Finishing, on the Most Approved English and French Methods:

Being Practical Instructions in Dyeing Silks, Woollens, and Cottons, Feathers, Chips, Straw, etc. Scouring and Cleaning Bed and Window Curtains, Carpets, Rugs, etc. French and English Cleaning, any Color or Fabric of Silk, Satin, or Damask. By THOMAS LOVE, a Working Dyer and Scourer. Second American Edition, to which are added General Instructions for the Use of Aniline Colors. In one volume, 8vo., 343 pages. $5.00

MAIN and BROWN.—Questions on Subjects Connected with the Marine Steam-Engine:

And Examination Papers; with Hints for their Solution. By THOMAS J. MAIN, Professor of Mathematics, Royal Naval College, and THOMAS BROWN, Chief Engineer, R. N. 12mo., cloth.   .   .   . $1.50

MAIN and BROWN.—The Indicator and Dynamometer:

With their Practical Applications to the Steam-Engine. By THOMAS J. MAIN, M. A. F. R., Assistant Professor Royal Naval College, Portsmouth, and THOMAS BROWN, Assoc. Inst. C. E., Chief Engineer, R. N., attached to the Royal Naval College. Illustrated. From the Fourth London Edition. 8vo. $1.50

MAIN and BROWN.—The Marine Steam-Engine.

By THOMAS J. MAIN, F. R.; Assistant S. Mathematical Professor at the Royal Naval College, Portsmouth, and, THOMAS BROWN, Assoc. Inst. C. E., Chief Engineer R. N. Attached to the Royal Naval College. Authors of "Questions connected with the Marine Steam-Engine," and the "Indicator and Dynamometer." With numerous Illustrations. In one volume, 8vo. $5.00

MARTIN.—Screw-Cutting Tables, for the Use of Mechanical Engineers:

Showing the Proper Arrangement of Wheels for Cutting the Threads of Screws of any required Pitch; with a Table for Making the Universal Gas-Pipe Thread and Taps. By W. A. MARTIN, Engineer. 8vo.   50

Mechanics' (Amateur) Workshop:

A treatise containing plain and concise directions for the manipulation of Wood and Metals, including Casting, Forging, Brazing, Soldering, and Carpentry. By the author of the "Lathe and its Uses." Third edition. Illustrated. 8vo. $3.00

MOLESWORTH.—Pocket-Book of Useful Formulæ and Memoranda for Civil and Mechanical Engineers.

By GUILFORD L. MOLESWORTH, Member of the Institution of Civil Engineers, Chief Resident Engineer of the Ceylon Railway. Second American, from the Tenth London Edition. In one volume, full bound in pocket-book form. $2.00

NAPIER.—A System of Chemistry Applied to Dyeing.

By JAMES NAPIER, F. C. S. A New and Thoroughly Revised Edition. Completely brought up to the present state of the Science, including the Chemistry of Coal Tar Colors, by A. A. FESQUET, Chemist and Engineer. With an Appendix on Dyeing and Calico Printing, as shown at the Universal Exposition, Paris, 1867. Illustrated. In one volume, 8vo., 422 pages. $5.00

NAPIER.—Manual of Electro-Metallurgy:

Including the Application of the Art to Manufacturing Processes. By JAMES NAPIER. Fourth American, from the Fourth London edition, revised and enlarged. Illustrated by engravings. In one vol., 8vo. $2.00

NASON.—Table of Reactions for Qualitative Chemical Analysis.

By HENRY B. NASON, Professor of Chemistry in the Rensselaer Polytechnic Institute, Troy, New York. Illustrated by Colors.   .   63

NEWBERY.—Gleanings from Ornamental Art of every style:

Drawn from Examples in the British, South Kensington, Indian, Crystal Palace, and other Museums, the Exhibitions of 1851 and 1862, and the best English and Foreign works. In a series of one hundred exquisitely drawn Plates, containing many hundred examples. By ROBERT NEWBERY. 4to. $15.00

NICHOLSON.—A Manual of the Art of Bookbinding:

Containing full instructions in the different Branches of Forwarding, Gilding, and Finishing. Also, the Art of Marbling Book-edges and Paper. By JAMES B. NICHOLSON. Illustrated. 12mo., cloth. $2.25

NICHOLSON.—The Carpenter's New Guide:

A Complete Book of Lines for Carpenters and Joiners. By PETER NICHOLSON. The whole carefully and thoroughly revised by H. K. DAVIS, and containing numerous new and improved and original Designs for Roofs, Domes, etc. By SAMUEL SLOAN, Architect. Illustrated by 80 plates. 4to. $4.50

NORRIS.—A Hand-book for Locomotive Engineers and Machinists:

Comprising the Proportions and Calculations for Constructing Locomotives; Manner of Setting Valves; Tables of Squares, Cubes, Areas, etc., etc. By SEPTIMUS NORRIS, Civil and Mechanical Engineer. New edition. Illustrated. 12mo., cloth. $2.00

NYSTROM.—On Technological Education, and the Construction of Ships and Screw Propellers:

For Naval and Marine Engineers. By JOHN W. NYSTROM, late Acting Chief Engineer, U. S. N. Second edition, revised with additional matter. Illustrated by seven engravings. 12mo.   .   . $1.50

O'NEILL.—A Dictionary of Dyeing and Calico Printing:

Containing a brief account of all the Substances and Processes in use in the Art of Dyeing and Printing Textile Fabrics; with Practical Receipts and Scientific Information. By CHARLES O'NEILL, Analytical Chemist; Fellow of the Chemical Society of London; Member of the Literary and Philosophical Society of Manchester; Author of "Chemistry of Calico Printing and Dyeing." To which is added an Essay on Coal Tar Colors and their application to Dyeing and Calico Printing. By A. A. FESQUET, Chemist and Engineer. With an Appendix on Dyeing and Calico Printing, as shown at the Universal Exposition, Paris, 1867. In one volume, 8vo., 491 pages.   . $6.00

ORTON.—Underground Treasures:

How and Where to Find Them. A Key for the Ready Determination of all the Useful Minerals within the United States. By JAMES ORTON, A. M. Illustrated, 12mo. $1.50

OSBORN.—American Mines and Mining:

Theoretically and Practically Considered. By Prof. H. S. OSBORN. Illustrated by numerous engravings. 8vo. (*In preparation.*)

OSBORN.—The Metallurgy of Iron and Steel:

Theoretical and Practical in all its Branches; with special reference to American Materials and Processes. By H. S. OSBORN, LL. D., Professor of Mining and Metallurgy in Lafayette College, Easton, Pennsylvania. Illustrated by numerous large folding plates and wood-engravings. 8vo. $15.00

OVERMAN.—The Manufacture of Steel:

Containing the Practice and Principles of Working and Making Steel. A Handbook for Blacksmiths and Workers in Steel and Iron, Wagon Makers, Die Sinkers, Cutlers, and Manufacturers of Files and Hardware, of Steel and Iron, and for Men of Science and Art. By FREDERICK OVERMAN, Mining Engineer, Author of the "Manufacture of Iron," etc. A new, enlarged, and revised Edition. By A. A. FESQUET, Chemist and Engineer. $1.50

OVERMAN.—The Moulder and Founder's Pocket Guide:

A Treatise on Moulding and Founding in Green-sand, Dry-sand, Loam, and Cement; the Moulding of Machine Frames, Mill-gear, Hollow-ware, Ornaments, Trinkets, Bells, and Statues; Description of Moulds for Iron, Bronze, Brass, and other Metals; Plaster of Paris, Sulphur, Wax, and other articles commonly used in Casting; the Construction of Melting Furnaces, the Melting and Founding of Metals; the Composition of Alloys and their Nature. With an Appendix containing Receipts for Alloys, Bronze, Varnishes and Colors for Castings; also, Tables on the Strength and other qualities of Cast Metals. By FREDERICK OVERMAN, Mining Engineer, Author of "The Manufacture of Iron." With 42 Illustrations. 12mo. $1.50

Painter, Gilder, and Varnisher's Companion:

Containing Rules and Regulations in everything relating to the Arts of Painting, Gilding, Varnishing, Glass-Staining, Graining, Marbling, Sign-Writing, Gilding on Glass, and Coach Painting and Varnishing; Tests for the Detection of Adulterations in Oils, Colors, etc.; and a Statement of the Diseases to which Painters are peculiarly liable, with the Simplest and Best Remedies. Sixteenth Edition. Revised, with an Appendix. Containing Colors and Coloring – Theoretical and Practical. Comprising descriptions of a great variety of Additional Pigments, their Qualities and Uses, to which are added, Dryers, and Modes and Operations of Painting, etc. Together with Chevreul's Principles of Harmony and Contrast of Colors. 12mo., cloth. $1.50

PALLETT.—The Miller's, Millwright's, and Engineer's Guide.

By HENRY PALLETT. Illustrated. In one volume, 12mo. $3.00

PERCY.—The Manufacture of Russian Sheet-Iron.

By JOHN PERCY, M.D., F.R.S., Lecturer on Metallurgy at the Royal School of Mines, and to The Advanced Class of Artillery Officers at the Royal Artillery Institution, Woolwich; Author of "Metallurgy." With Illustrations. 8vo., paper. . . . . . . 50 cts.

PERKINS.—Gas and Ventilation.

Practical Treatise on Gas and Ventilation. With Special Relation to Illuminating, Heating, and Cooking by Gas. Including Scientific Helps to Engineer-students and others. With Illustrated Diagrams. By E. E. PERKINS. 12mo., cloth. $1.25

PERKINS and STOWE.—A New Guide to the Sheet-iron and Boiler Plate Roller:

Containing a Series of Tables showing the Weight of Slabs and Piles to produce Boiler Plates, and of the Weight of Piles and the Sizes of Bars to produce Sheet-iron; the Thickness of the Bar Gauge in decimals; the Weight per foot, and the Thickness on the Bar or Wire Gauge of the fractional parts of an inch; the Weight per sheet, and the Thickness on the Wire Gauge of Sheet-iron of various dimensions to weigh 112 lbs. per bundle; and the conversion of Short Weight into Long Weight, and Long Weight into Short. Estimated and collected by G. H. PERKINS and J. G. STOWE. $2.50

PHILLIPS and DARLINGTON.—Records of Mining and Metallurgy;

Or Facts and Memoranda for the use of the Mine Agent and Smelter. By J. ARTHUR PHILLIPS, Mining Engineer, Graduate of the Imperial School of Mines, France, etc., and JOHN DARLINGTON. Illustrated by numerous engravings. In one volume, 12mo. . . $2.00

PROTEAUX.—Practical Guide for the Manufacture of Paper and Boards.

By A. PROTEAUX, Civil Engineer, and Graduate of the School of Arts and Manufactures, and Director of Thiers' Paper Mill, Puy-de-Dôme. With additions, by L. S. LE NORMAND. Translated from the French, with Notes, by HORATIO PAINE, A. B., M. D. To which is added a Chapter on the Manufacture of Paper from Wood in the United States, by HENRY T. BROWN, of the "American Artisan." Illustrated by six plates, containing Drawings of Raw Materials, Machinery, Plans of Paper-Mills, etc., etc. 8vo. $7.50

REGNAULT.—Elements of Chemistry.

By M. V. REGNAULT. Translated from the French by T. FORREST BETTON, M. D., and edited, with Notes, by JAMES C. BOOTH, Melter and Refiner U. S. Mint, and WM. L. FABER, Metallurgist and Mining Engineer. Illustrated by nearly 700 wood engravings. Comprising nearly 1500 pages. In two volumes, 8vo., cloth. . . . $7.50

REID.—A Practical Treatise on the Manufacture of Portland Cement:

By HENRY REID, C. E. To which is added a Translation of M. A. Lipowitz's Work, describing a New Method adopted in Germany for Manufacturing that Cement, by W. F. REID. Illustrated by plates and wood engravings. 8vo. $6.00

RIFFAULT, VERGNAUD, and TOUSSAINT.—A Practical Treatise on the Manufacture of Varnishes.

By M M. RIFFAULT, VERGNAUD, and TOUSSAINT. Revised and Edited by M. F. MALEPEYRE and Dr. EMIL WINCKLER. Illustrated. In one volume, 8vo. (*In preparation.*)

RIFFAULT, VERGNAUD, and TOUSSAINT.—A Practical Treatise on the Manufacture of Colors for Painting:

Containing the best Formulæ and the Processes the Newest and in most General Use. By M M. RIFFAULT, VERGNAUD, and TOUSSAINT. Revised and Edited by M. F. MALEPEYRE and Dr. EMIL WINCKLER. Translated from the French by A. A. FESQUET, Chemist and Engineer. Illustrated by Engravings. In one volume, 650 pages, 8vo. (*Ready June 1, 1874.*)

ROBINSON.—Explosions of Steam Boilers:

How they are Caused, and how they may be Prevented. By J. R. ROBINSON, Steam Engineer. 12mo. $1.25

ROPER.—A Catechism of High Pressure or Non-Condensing Steam-Engines:

Including the Modelling, Constructing, Running, and Management of Steam Engines and Steam Boilers. With Illustrations. By STEPHEN ROPER, Engineer. Full bound tucks . . . $2.00

ROSELEUR.—Galvanoplastic Manipulations:

A Practical Guide for the Gold and Silver Electro-plater and the Galvanoplastic Operator. Translated from the French of ALFRED ROSELEUR, Chemist, Professor of the Galvanoplastic Art, Manufacturer of Chemicals, Gold and Silver Electro-plater. By A. A. FESQUET, Chemist and Engineer. Illustrated by over 127 Engravings on wood. 8vo., 495 pages. $6.00

☞ *This Treatise is the fullest and by far the best on this subject ever published in the United States.*

SCHINZ.—Researches on the Action of the Blast Furnace.

By CHARLES SCHINZ. Translated from the German with the special permission of the Author by WILLIAM H. MAW and MORITZ MULLER. With an Appendix written by the Author expressly for this edition. Illustrated by seven plates, containing 28 figures. In one volume, 12mo. $4.25

SHAW.—Civil Architecture:

Being a Complete Theoretical and Practical System of Building, containing the Fundamental Principles of the Art. By EDWARD SHAW, Architect. To which is added a Treatise on Gothic Architecture, etc. By THOMAS W. SILLOWAY and GEORGE M. HARDING, Architects. The whole illustrated by One Hundred and Two quarto plates finely engraved on copper. Eleventh Edition. 4to., cloth. . $10.00

SHUNK.—A Practical Treatise on Railway Curves and Location, for Young Engineers.

By WILLIAM F. SHUNK, Civil Engineer. 12mo. . . $2.00

SLOAN.—American Houses:

A variety of Original Designs for Rural Buildings. Illustrated by 26 colored Engravings, with Descriptive References. By SAMUEL SLOAN, Architect, author of the "Model Architect," etc., etc. 8vo. $2.50

SMEATON.—Builder's Pocket Companion:

Containing the Elements of Building, Surveying, and Architecture; with Practical Rules and Instructions connected with the subject. By A. C. SMEATON, Civil Engineer, etc. In one volume, 12mo. $1.50

SMITH.—A Manual of Political Economy.

By E. PESHINE SMITH. A new Edition, to which is added a full Index. 12mo., cloth. $1.25

SMITH.—Parks and Pleasure Grounds:

Or Practical Notes on Country Residences, Villas, Public Parks, and Gardens. By CHARLES H. J. SMITH, Landscape Gardener and Garden Architect, etc., etc. 12mo. $2.25

SMITH.—The Dyer's Instructor:

Comprising Practical Instructions in the Art of Dyeing Silk, Cotton, Wool, and Worsted, and Woollen Goods: containing nearly 800 Receipts. To which is added a Treatise on the Art of Padding; and the Printing of Silk Warps, Skeins, and Handkerchiefs, and the various Mordants and Colors for the different styles of such work. By DAVID SMITH, Pattern Dyer. 12mo., cloth. . . . $3.00

SMITH.—The Practical Dyer's Guide:

Comprising Practical Instructions in the Dyeing of Shot Cobourgs, Silk Striped Orleans, Colored Orleans from Black Warps, Ditto from White Warps, Colored Cobourgs from White Warps, Merinos, Yarns, Woollen Cloths, etc. Containing nearly 300 Receipts, to most of which a Dyed Pattern is annexed. Also, A Treatise on the Art of Padding. By DAVID SMITH. In one volume, 8vo. Price. . . $25.00

STEWART.—The American System.

Speeches on the Tariff Question, and on Internal Improvements, principally delivered in the House of Representatives of the United States. By ANDREW STEWART, late M. C. from Pennsylvania. With a Portrait, and a Biographical Sketch. In one volume, 8vo., 407 pages. $3.00

STOKES.—Cabinet-maker's and Upholsterer's Companion:

Comprising the Rudiments and Principles of Cabinet-making and Upholstery, with Familiar Instructions, illustrated by Examples for attaining a Proficiency in the Art of Drawing, as applicable to Cabinet-work; the Processes of Veneering, Inlaying, and Buhl-work; the Art of Dyeing and Staining Wood, Bone, Tortoise Shell, etc. Directions for Lackering, Japanning, and Varnishing; to make French Polish; to prepare the Best Glues, Cements, and Compositions, and a number of Receipts particularly useful for workmen generally. By J. STOKES. In one volume, 12mo. With Illustrations.   . $1.25

Strength and other Properties of Metals:

Reports of Experiments on the Strength and other Properties of Metals for Cannon. With a Description of the Machines for testing Metals, and of the Classification of Cannon in service. By Officers of the Ordnance Department U. S. Army. By authority of the Secretary of War. Illustrated by 25 large steel plates. In one volume, 4to.   . $10.00

SULLIVAN.—Protection to Native Industry.

By Sir EDWARD SULLIVAN, Baronet, author of "Ten Chapters on Social Reforms." In one volume, 8vo. $1.50

Tables Showing the Weight of Round, Square, and Flat Bar Iron, Steel, etc.,

By Measurement. Cloth.   63

TAYLOR.—Statistics of Coal:

Including Mineral Bituminous Substances employed in Arts and Manufactures; with their Geographical, Geological, and Commercial Distribution and Amount of Production and Consumption on the American Continent. With Incidental Statistics of the Iron Manufacture. By R. C. TAYLOR. Second edition, revised by S. S. HALDEMAN. Illustrated by five Maps and many wood engravings. 8vo., cloth. $10.00

TEMPLETON.—The Practical Examinator on Steam and the Steam-Engine:

With Instructive References relative thereto, arranged for the Use of Engineers, Students, and others. By WM. TEMPLETON, Engineer. 12mo. $1.25

THOMAS.—The Modern Practice of Photography.

By R. W. THOMAS, F. C. S. 8vo., cloth.   75

THOMSON.—Freight Charges Calculator.

By ANDREW THOMSON, Freight Agent. 24mo. .   .   . $1.25

TURNING: Specimens of Fancy Turning Executed on the Hand or Foot Lathe:

With Geometric, Oval, and Eccentric Chucks, and Elliptical Cutting Frame. By an Amateur. Illustrated by 30 exquisite Photographs. 4to. $3.00

Turner's (The) Companion:

Containing Instructions in Concentric, Elliptic, and Eccentric Turning: also various Plates of Chucks, Tools, and Instruments; and Directions for using the Eccentric Cutter, Drill, Vertical Cutter, and Circular Rest; with Patterns and Instructions for working them. A new edition in one volume, 12mo. $1.50

URBIN.—BRULL.—A Practical Guide for Puddling Iron and Steel.

By ED. URBIN, Engineer of Arts and Manufactures. A Prize Essay read before the Association of Engineers, Graduate of the School of Mines, of Liege, Belgium, at the Meeting of 1865–6. To which is added A COMPARISON OF THE RESISTING PROPERTIES OF IRON AND STEEL. By A. BRULL. Translated from the French by A. A. FESQUET, Chemist and Engineer. In one volume, 8vo. $1.00

VAILE.—Galvanized Iron Cornice-Worker's Manual:

Containing Instructions in Laying out the Different Mitres, and Making Patterns for all kinds of Plain and Circular Work. Also, Tables of Weights, Areas and Circumferences of Circles, and other Matter calculated to Benefit the Trade. By CHARLES A. VAILE, Superintendent "Richmond Cornice Works," Richmond, Indiana. Illustrated by 21 Plates. In one volume, 4to. $5.00

VILLE.—The School of Chemical Manures:

Or, Elementary Principles in the Use of Fertilizing Agents. From the French of M. GEORGE VILLE, by A. A. FESQUET, Chemist and Engineer. With Illustrations. In one volume, 12 mo. . . $1.25

VOGDES.—The Architect's and Builder's Pocket Companion and Price Book:

Consisting of a Short but Comprehensive Epitome of Decimals, Duodecimals, Geometry and Mensuration; with Tables of U. S. Measures, Sizes, Weights, Strengths, etc., of Iron, Wood, Stone, and various other Materials, Quantities of Materials in Given Sizes, and Dimensions of Wood, Brick, and Stone; and a full and complete Bill of Prices for Carpenter's Work; also, Rules for Computing and Valuing Brick and Brick Work, Stone Work, Painting, Plastering, etc. By FRANK W. VOGDES, Architect. Illustrated. Full bound in pocket-book form. $2.00
Bound in cloth. 1.50

WARN.—The Sheet-Metal Worker's Instructor:

For Zinc, Sheet-Iron, Copper, and Tin-Plate Workers, etc. Containing a selection of Geometrical Problems; also, Practical and Simple Rules for describing the various Patterns required in the different branches of the above Trades. By REUBEN H. WARN, Practical Tin-plate Worker. To which is added an Appendix, containing Instructions for Boiler Making, Mensuration of Surfaces and Solids, Rules for Calculating the Weights of different Figures of Iron and Steel, Tables of the Weights of Iron, Steel, etc. Illustrated by 32 Plates and 37 Wood Engravings. 8vo. $3.00

WARNER.—New Theorems, Tables, and Diagrams for the Computation of Earth-Work:

Designed for the use of Engineers in Preliminary and Final Estimates, of Students in Engineering, and of Contractors and other non-professional Computers. In Two Parts, with an Appendix. Part I.—A Practical Treatise; Part II.—A Theoretical Treatise; and the Appendix. Containing Notes to the Rules and Examples of Part I.; Explanations of the Construction of Scales, Tables, and Diagrams, and a Treatise upon Equivalent Square Bases and Equivalent Level Heights. The whole illustrated by numerous original Engravings, comprising Explanatory Cuts for Definitions and Problems, Stereometric Scales and Diagrams, and a Series of Lithographic Drawings from Models, showing all the Combinations of Solid Forms which occur in Railroad Excavations and Embankments. By JOHN WARNER, A. M., Mining and Mechanical Engineer. 8vo. $5.00

WATSON.—A Manual of the Hand-Lathe:

Comprising Concise Directions for working Metals of all kinds, Ivory, Bone and Precious Woods; Dyeing, Coloring, and French Polishing; Inlaying by Veneers, and various methods practised to produce Elaborate work with Dispatch, and at Small Expense. By EGBERT P. WATSON, late of "The Scientific American," Author of "The Modern Practice of American Machinists and Engineers." Illustrated by 78 Engravings. $1.50

WATSON.—The Modern Practice of American Machinists and Engineers:

Including the Construction, Application, and Use of Drills, Lathe Tools, Cutters for Boring Cylinders, and Hollow Work Generally, with the most Economical Speed for the same; the Results verified by Actual Practice at the Lathe, the Vice, and on the Floor. Together with Workshop Management, Economy of Manufacture, the Steam-Engine, Boilers, Gears, Belting, etc., etc. By EGBERT P. WATSON, late of the "Scientific American." Illustrated by 86 Engravings. In one volume, 12mo. $2.50

WATSON.—The Theory and Practice of the Art of Weaving by Hand and Power:

With Calculations and Tables for the use of those connected with the Trade. By JOHN WATSON, Manufacturer and Practical Machine Maker. Illustrated by large Drawings of the best Power Looms. 8vo. $10.00

WEATHERLY.—Treatise on the Art of Boiling Sugar, Crystallizing, Lozenge-making, Comfits, Gum Goods.

12mo. $2.00

WEDDING.—The Metallurgy of Iron;

Theoretically and Practically Considered. By Dr. HERMANN WEDDING, Professor of the Metallurgy of Iron at the Royal Mining Academy, Berlin. Translated by JULIUS DU MONT, Bethlehem, Pa. Illustrated by 207 Engravings on Wood, and three Plates. In one volume, 8vo. (*In press.*)

WILL.—Tables for Qualitative Chemical Analysis.

By Professor HEINRICH WILL, of Giessen, Germany. Seventh edition. Translated by CHARLES F. HIMES, Ph. D., Professor of Natural Science, Dickinson College, Carlisle, Pa. . . .

WILLIAMS.—On Heat and Steam :

Embracing New Views of Vaporization, Condensation, and Explosions. By CHARLES WYE WILLIAMS, A. I. C. E. Illustrated. 8vo. $3.50

WOHLER.—A Hand-Book of Mineral Analysis.

By F. WOHLER, Professor of Chemistry in the University of Göttingen. Edited by HENRY B. NASON, Professor of Chemistry in the Rensselaer Polytechnic Institute, Troy, New York. Illustrated. In one volume, 12mo. $3 00

WORSSAM.—On Mechanical Saws :

From the Transactions of the Society of Engineers, 1869. By S. W. WORSSAM, Jr. Illustrated by 18 large plates. 8vo. . . $5.00

767171

Printed in Great Britain by
Amazon.co.uk, Ltd.,
Marston Gate.